AF447782

*A Text Book of*

# EMERGING TRENDS IN MECHANICAL ENGINEERING

## (22652)

## Semester – VI

### THRIED YEAR DIPLOMA IN MECHANICAL ENGINEERING GROUP

*As Per MSBTE's 'I' Scheme Syllabus*


**MR. MD. MAROOF MD. NAYEEM**

ME (Manufacturing Engg.), BE (Mech. Engg.)
LMISTE, IAENGG, IFERP, ISRD, IAER,
Lecturer, Mechanical Engineering Department,
Anjuman-I-Islam's A.R. Kalsekar Polytechnic,
Panvel.

**MR. ADNAN IQBAL SHAIKH**

ME (Machine Design), BE (Mech. Engg.)
DME, AMIE, LMISTE,
Lecturer, Mechanical Engineering Department,
Anjuman-I-Islam's A.R. Kalsekar Polytechnic,
Panvel.

**MR. SANJAY S. NIBANDHE**

M.S in Manufacturing Management, MMS
Member in Syllabus Formation of D. Voc, B. Voc in
Automobile Courses, Industry Experts for
Vehicle Design Development and Manufacturing Evaluation
35 Plus Years in 2 Wheeler, 4 Wheeler & Commercial Vehicle Development

N4564

**Emerging Trends in Mechanical Engineering (Sem. – VI)**      **ISBN 978-93-89825-84-8**

**First Edition**   **:**   **January 2020**

©        **:**   **Authors**


The text of this publication, or any part thereof, should not be reproduced or transmitted in any form or stored in any computer storage system or device for distribution including photocopy, recording, taping or information retrieval system or reproduced on any disc, tape, perforated media or other information storage device etc., without the written permission of Authors with whom the rights are reserved. Breach of this condition is liable for legal action.

Every effort has been made to avoid errors or omissions in this publication. In spite of this, errors may have crept in. Any mistake, error or discrepancy so noted and shall be brought to our notice shall be taken care of in the next edition. It is notified that neither the publisher nor the authors or seller shall be responsible for any damage or loss of action to any one, of any kind, in any manner, therefrom.

**Published By :**

**NIRALI PRAKASHAN**

Abhyudaya Pragati, 1312, Shivaji Nagar

Off J.M. Road, PUNE – 411005

Tel - (020) 25512336/37/39, Fax - (020) 25511379

Email : niralipune@pragationline.com

➢ **DISTRIBUTION CENTRES**

**PUNE**

**Nirali Prakashan**   **:**   119, Budhwar Peth, Jogeshwari Mandir Lane, Pune 411002, Maharashtra

(For orders within Pune)     Tel : (020) 2445 2044, Fax : (020) 2445 1538; Mobile : 9657703145

Email : bookorder@pragationline.com, niralilocal@pragationline.com

**Nirali Prakashan**   **:**   S. No. 28/27, Dhyari, Near Pari Company, Pune 411041

(For orders outside Pune)     Tel : (020) 24690204 Fax : (020) 24690316; Mobile : 9657703143

Email : dhyari@pragationline.com, bookorder@pragationline.com

**MUMBAI**

**Nirali Prakashan**   **:**   385, S.V.P. Road, Rasdhara Co-op. Hsg. Society Ltd.,

Girgaum, Mumbai 400004, Maharashtra; Mobile : 9320129587

Tel : (022) 2385 6339 / 2386 9976, Fax : (022) 2386 9976

Email : niralimumbai@pragationline.com

➢ **DISTRIBUTION BRANCHES**

**JALGAON**

**Nirali Prakashan**   **:**   34, V. V. Golani Market, Navi Peth, Jalgaon 425001,

Maharashtra, Tel : (0257) 222 0395, Mob : 94234 91860

Email : niralijalgoan@pragationline.com

**KOLHAPUR**

**Nirali Prakashan**   **:**   New Mahadvar Road, Kedar Plaza, 1st Floor Opp. IDBI Bank

Kolhapur 416 012, Maharashtra. Mob : 9850046155

Email : niralikolhapur@pragationline.com

**NAGPUR**

**Nirali Prakashan**   **:**   Above Maratha Mandir, Shop No. 3, First Floor,

Rani Jhanshi Square, Sitabuldi, Nagpur 440012, Maharashtra

Tel : (0712) 254 7129; Email : niralinagpur@pragationline.com

**DELHI**

**Nirali Prakashan**   **:**   4593/15, Basement, Agarwal Lane, Ansari Road, Daryaganj

Near Times of India Building, New Delhi  110002 Mob :  08505972553

Email : niralidelhi@pragationline.com

**BANGALURU**

**Nirali Prakashan**   **:**   Maitri Ground Floor, Jaya Apartments, No. 99, 6th Cross, 6th Main,

Malleswaram, Bangaluru  560 003, Karnataka

Mob : +91 9449043034

Email: niralibangalore@pragationline.com

**Note :** Every possible effort has been made to avoid errors or omissions in this book. In spite this, errors may have crept in. Any type of error or mistake so noted, and shall be brought to our notice, shall be taken care of in the next edition. It is notified that neither the publisher, nor the author or book seller shall be responsible for any damage or loss of action to any one of any kind, in any manner, therefrom. The reader must cross check all the facts and contents with original Government notification or publications.

niralipune@pragationline.com  |  www.pragationline.com

Also find us on   www.facebook.com/niralibooks

# Preface ...

It gives us a great pleasure to present this textbook on "Emerging Trends in Mechanical Engineering" for the students of Third Year Diploma courses in Mechanical Engineering. This book covers the complete revised syllabus prescribed by the Board of Technical Examination, Maharashtra State. The need of such book is felt due to fact that the students, at large, find themselves confused and lost, while referring so on any books available on ""Emerging Trends in Mechanical Engineering"". Therefore an attempt has been made by us to fulfill the students in a single book.

In this book, both theoretical concept and numerical illustrations have been appropriately balanced to provide clarity of the subject matter. Typical questions have been added at the end of each chapter to enhance the utility of this book.

The authors express their sincere thanks to publisher Shri Dineshbhai Furia, Shri Jignesh Furia and all the staff of Nirali Prakashan, especially Mr. Ilyas Shaikh, Ms. Anagha Medhekar and Ms. Yojana Deshpande, Mr. Ravindra Walodare for their non-stop efforts for making this book excellent.

It will not be out of place here to express thanks to our family members for extending their support, co-operation and help during the writing of this book.

The constructive suggestions and criticism will be greatly appreciated for enhancing the utility of the book.

**18th January 2020**                                                                                   **Authors**

## 1. Recent Trends in Automobile Industry　　　(Marks 20 - Hours 14)

1.1　Hybrid cars-manufacturers, Types - Micro hybrid, Mild hybrid, Full hybrid, Series hybrid, Parallel hybrid.

1.2　E-vehicles - Manufacturers, Specifications, Types of batteries, LI-ion batteries, Sodium nickel chloride batteries, Sodium sulphur batteries, Fuel cell. Charging - Charging methods and modes. Issues with E-vehicles.

1.3　Safety in Automobile - Air bags, Automatic emergency braking, Adaptive cruise control, Electronic stability programmer, Anti-collision system, Active passive integration system.

## 2. Process Engineering　　　(Marks 10 - Hours 06)

2.1.　Process Boiler, Steam, Condensate loop in process industries.

2.2.　Introduction to Ultra Supper critical Boiler, Principle, Working, advantages, application.

2.3.　Hyperbolic Cooling Towers.

2.4.　Waste heat recovery in process industries.

## 3. Recent Trends in Manufacturing Engineering　　　(Marks 20 - Hours 14)

3.1　**Smart Manufacturing Technology:** Introduction, Elements and Applications.

3.2　**Automation:** Need, Basic elements of automated systems, Automation principles and strategies, Benefits.

3.3　**Types of Automation:** Fixed, Programmable, Flexible, Hard and soft automation.

3.4　**Industrial Robotics:** Robot anatomy, Robot control systems, End effectors, Sensors in robotics, Industrial robot applications.

3.5　4-D Printing Technology - Printing Techniques, 3D scanning technology - function, Applications.

## 4. Energy Monitoring, Management and Audit　　　(Marks 10 - Hours 08)

4.1　Standards and labeling Standards (HVAC)

4.2　**Energy Monitoring and Targeting:** Defining monitoring and targeting, Elements of monitoring and targeting, Methods Monitoring and Targeting System

4.3　**Energy Management and Audit:** Definition, Energy audit- need, Types of energy audit, Energy management (Audit) approach-understanding energy costs.

## 5. Agriculture Equipment and Post-Harvest Technology　　　(Marks 10 - Hours 06)

5.1　Tillers, Sowing and planting equipment, Welding machines, Spraying machines, Harvesting, Post harvesting machineries.

5.2　Elements of cold chain.

5.3　National Cooling Action Plan (NCAP)

# Contents ...

# Chapter 1

# RECENT TRENDS IN AUTOMOBILE INDUSTRY

**Weightage of Marks = 20, Teaching Hours = 14**

## Syllabus

1.1 Hybrid cars-manufacturers, Types - Micro hybrid, Mild hybrid, Full hybrid, Series hybrid, Parallel hybrid.

1.2 E-vehicles - Manufacturers, Specifications, Types of batteries, LI-ion batteries, Sodium nickel chloride batteries, Sodium sulphur batteries, Fuel cell. Charging - Charging methods and modes. Issues with e-vehicles.

1.3 Safety in Automobile - Air bags, Automatic emergency braking, Adaptive cruise control, Electronic stability programmer, Anti-collision system, Active passive integration system.

## About this Chapter

At the end of this chapter, students will be able to:

- Classify hybrid cars.
- List different batteries used in E-vehicles.
- Name different safety systems used in given vehicle.

## 1.1 HYBRID CARS MANUFACTURING

### 1.1.1 Introduction

The automobile Industry is being running for past 6 decades in full swings with IC engines. Over a period of time, fuel availability and emission becomes serious concerns.

The trend in India are also moved in similar ways. Slowly those norms for emission are enforced for all manufactures in past twenty years.

Current norms of Bharat Stage IV will get upgrade to Bharat Stage VI (BS-VI) and all the manufactures will make those changes in IC engines and try to meet those emission norms. However fossil fuels availability is big concern and increasing rates of imported crude oils is increasing burden to Indian economy with increase in demands.

Therefore, alternate fuels are being worked out and as out cum Bio-diesel, Ethanol, Methanol, Hydrogen etc. all those fuels are being tried out. The increasing demands are not yet met with those alternate fuels. In the meantime, new options of electric vehicles and Hybrid vehicle are experimented and being explored in Indian and Global market.

Hybrid Vehicles are mainly duel fuel vehicles for example gasoline and electric energy is used as source for automobile cars and buses. For low speed, low range city drive electrical energy is used.

Thereafter for highways and high speed gasoline fuel is used for efficient drives. It's understood that combination of two different fuels used for driving cars or buses needs different systems and mechanism for establishing drives.

The system become more complex however it delivers good mileage with low pollution in city area and speed at highways to cover distance very fast.

## 1.1.2 Concepts of Hybrid Electric Drive Trains

Normally any car need to develop sufficient power to meet the vehicle performance demands, adequate energy to support vehicle driving range, deliver high efficiency and emit low pollutants. Whenever more than one power trains are deployed, it's named as "Hybrid" vehicles.

The power train means power source or energy converter such as gasoline or diesel or hydrogen fuel cell or electric motor systems so on. A vehicle that has more than two power trains are called as Hybrid vehicles. A hybrid vehicle with an electric power train is called as HEV.

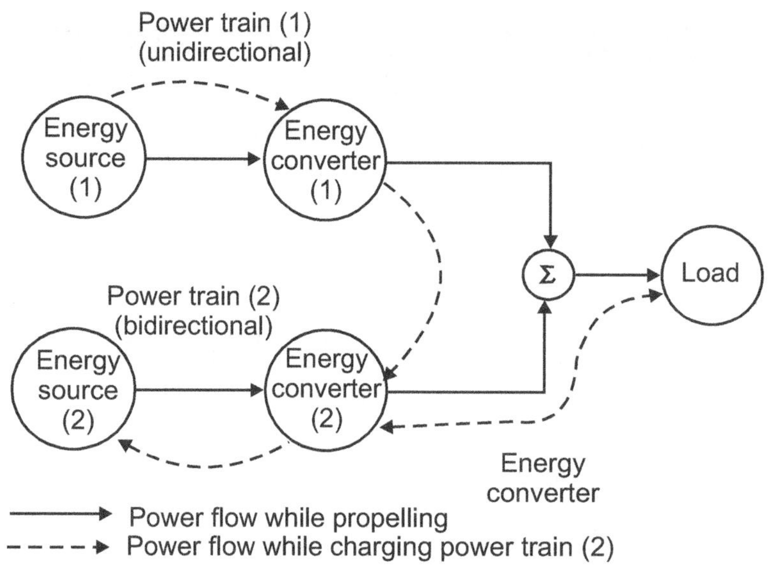


**Fig. 1.1: Conceptual Illustration of Hybrid Electric Drive Train**

More than two power train makes it complex drive train. IC engine plus electric motor system is explained in diagram (Fig. 1.1). To recapture the braking energy that is dissipated in the from of heat energy from IC engine.

In hybrid vehicles it is flow of energy in bidirectional or unidirectional. The figure explains possible power flow routes.

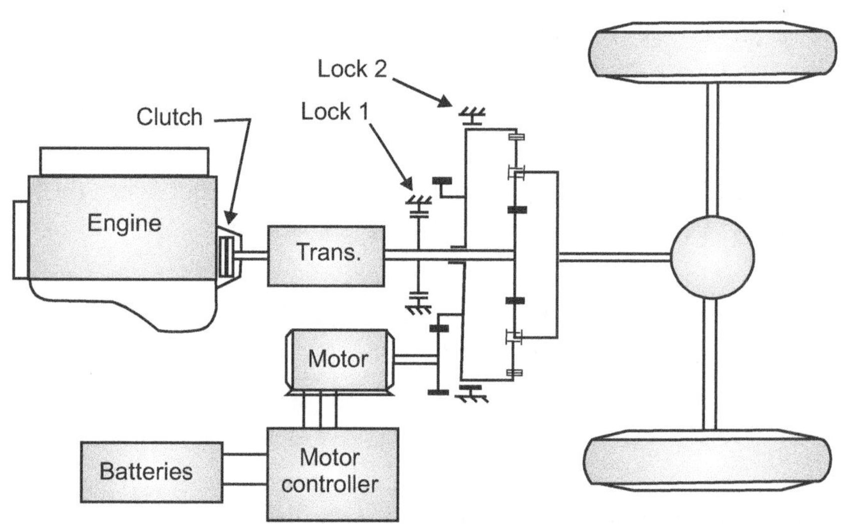


**Fig. 1.2: Hybrid Electric Drive Train with Speed Coupling of a Planetary Gear Unit**

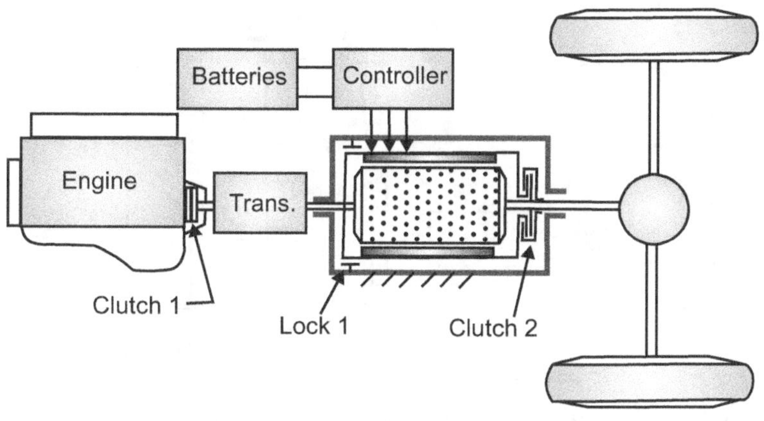


**Fig. 1.3: Hybrid Electric Drive Train with Speed Coupling of an Electric Transmotor**

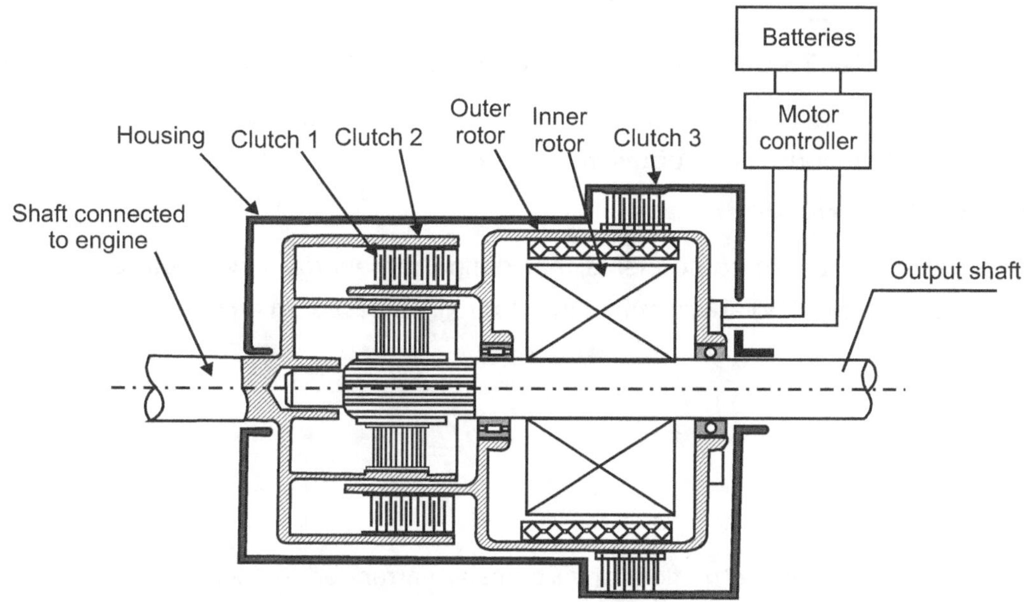


**Fig. 1.4: Implementation of Speed Coupling with a Transmotor**

A hybrid power train can supply its power to the load by selective power train. The various available patterns are as follows :

(i)   Power train 1 alone delivers power to load.

(ii)  Power train 2 alone deliver power to load.

(iii) Both power train 1 and 2 deliver power to load simultaneously.

(iv)  Power train 2 get power from 1 and then supply to load or vis-à-vis.

(v)   Regenerative power from breaking, feed to one unit.

(vi)  Power train 1 delivers to load and load deliver to power train 2.

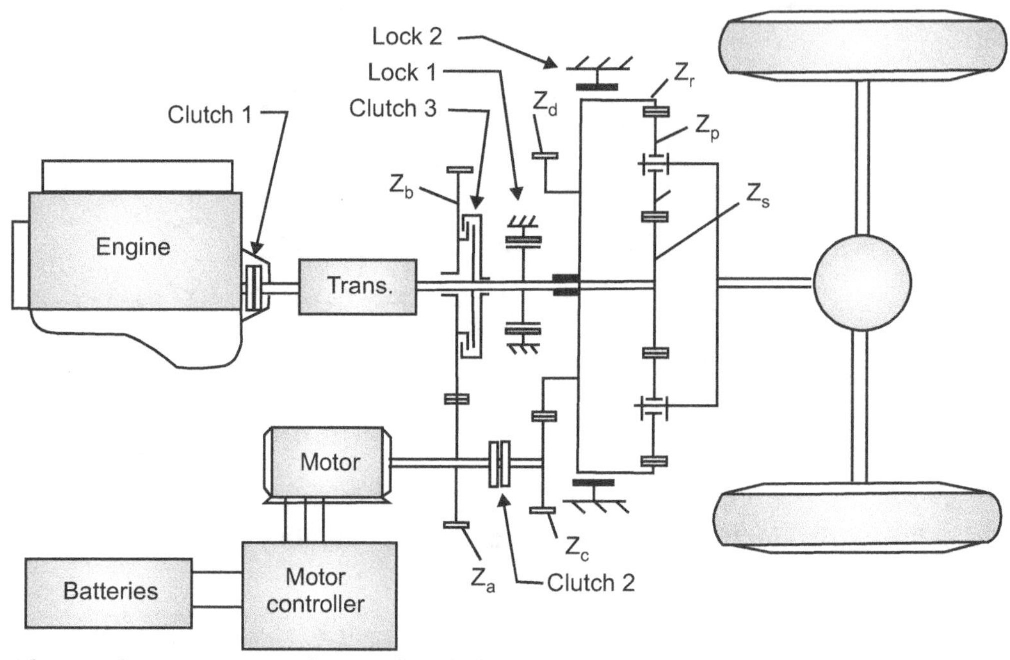


**Fig. 1.5: Alternative Torque and Speed Hybrid Electric Drive Train with a Planetary Gear Unit**

In this way various combinations of propelling from IC engine or Battery can be worked out. Thus the driving mode and the battery charging mode is also obtained in the effect.

### 1.1.3 Architecture of Hybrid Vehicles

In Hybrid the connection between components that define the energy flow route and control ports are key elements. Hybrid Electric Vehicles (HEV) are classified into four kinds based on architecture

   (a)  Series Hybrid

   (b)  Parallel Hybrid

   (c)  Series-Parallel Hybrid and

   (d)  Complex Hybrid.

In HEV mechanical and electrical energy flow drive trains are arranged in specific manners. HEV are defined on architecture based on power coupling or decoupling features such as electrical coupling, mechanical coupling and mechanical-electrical coupling drive trains.

**(a) Series Hybrid:** IC engine and electric motor connected in series its known as series hybrid and only electrical motor is providing power to wheels. The schematic diagram shows in figure (Fig. 1.6).

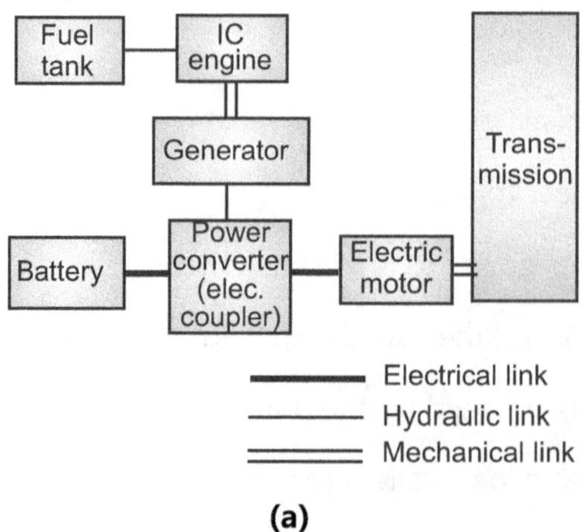


**(a)**

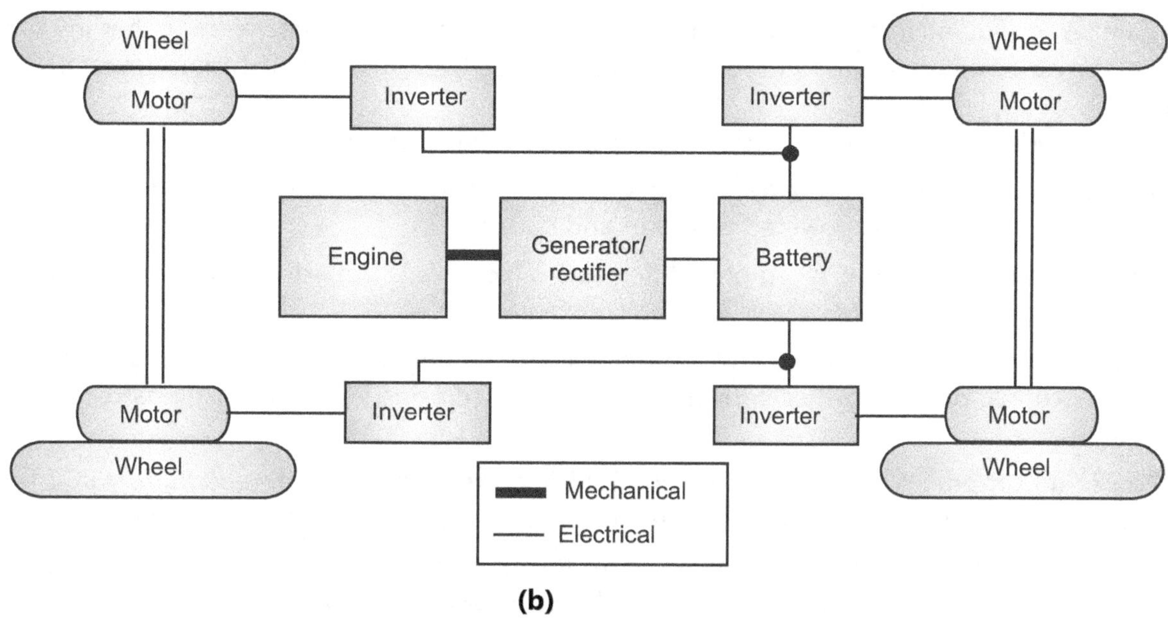



**(b)**

**Fig. 1.6: Hub Motor Configuration of a Series HEV**

In this HEV the IC engine is main energy source and converts its original energy from gasoline into mechanical power. The mechanical output of the ICE is converted into electrical energy using the generator.

The electric motor moves the final drive using electricity generated by generator or from energy stored in battery. The electric motor receives electric energy from engine through generator, or from battery or from both. Since engine is decoupled from wheels, the engine speed can be controlled independently of vehicle speed.

This is not only simplifying control mechanism but more importantly can allow engine operations independently to best optimum speed and fuel efficient way. The freedom to place engine provides flexibility at suitable position of vehicle. Generally, transmission is eliminated in series hybrid vehicles. The different combinations of propulsion are as follows-

- **Battery alone:**  Vehicle is powered by battery alone and IC engine is off during battery drive.  Sufficient electric charge is available with battery. Low speed operations in city area.  Advantage is no emission from IC engine.

- **Combined operation:** Whenever demand for speed and load increases it operates through ICE and Generator along with battery supply to electric motor.

- **Engine alone:** Vehicle is powered by only IC Engine during highway applications such as cruising mode or high speed runs. Battery supply is normally cut-off and neither its charged or discharge due to its state of charge (SOC).

- **Power Split:** Whenever ICE and Generator is on, vehicle power demand is blow optimum power and battery is at low SOC then part of generated power is used to charge battery.

- **Stationary charging:** The battery is getting charged without vehicle is being driven.

- **Regenerative braking:** The electric motor is operated as generator to convert the braking energy (kinetic) into electric energy and charge the battery.

In series HEV various configurations are used for efficient drives of vehicles. Number of electric motors increases from one to two or four. Those electric motors are placed near wheels to establish the direct drives to wheels.

**(b) Parallel Hybrid:** The IC Engine and electric motor are connected in parallel its known as Parallel HEV (PHEV).  The electricity stored in battery is recharged from electric grid. The ICE and Electric motor are coupled to final drive shaft through mechanical coupling and commonly used mechanisms such as clutch, gearbox belt drives or pulleys are used in it's configuration.

The parallel configuration allows both ICE and electric motor to drive the vehicle either in combined mode or separately. The electric motor is also used to absorb regenerative braking and capture excessive energy from ICE during cost down application.

The PHEV needs two propulsion devices. The various configurations are as follows:

- **Motor alone mode:** When battery is adequately charged and vehicle power demands low then engine is turned-off and vehicle is powered by motor and battery.

- **Combined power mode:** When power demand is high ICE is turned-on and motor also supplies power to wheels.

- **Engine alone mode:** When the cruising mode is used on highway and demand for power is relatively high, the ICE supply power to wheels for drive. The motor remains idle and mostly battery SOC is high level of charge.

- **Power split mode:** When engine is on and vehicle power demand is low and battery SOC is also at low level then portion of engine power is converted to electricity by motor to charge the battery.

- **Stationary charging mode:** The engine drives the electric motor as generator of power and then it charges the battery, however vehicle is not driven and remains as stationary.

- **Regenerative braking mode:** The electric motor is operated as generator to convert vehicle kinetic energy from braking into electric energy and charge the battery. Regenerative mode its possible to run ICE as well and provide additional current to battery for charging.

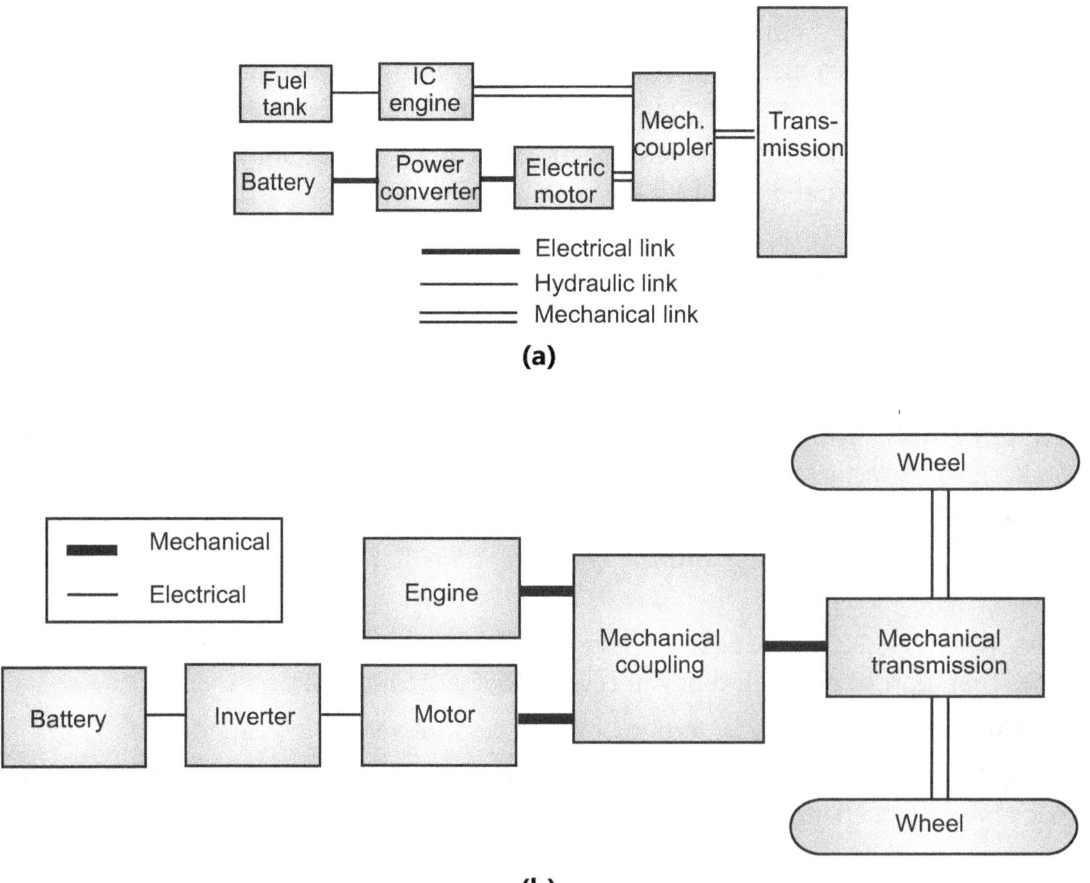


**Fig. 1.7: The Architecture of a Parallel HEV**

**(c) Series - Parallel Hybrid:** This type of hybrid are having features of both series and parallel HEV, it can be operated as series or parallel. In this concept engine is linked with final drive and therefore it can power to wheels directly. This works inline with series HEV.

Similarly, it provides second electric motor that serves as generator. Thus it can be comparable as parallel HEV. Since the series and parallel HEV operate it in both versions it delivers fuel efficiency along with excellent drivability.

Based on the demand, optimized performance can be obtained from this HEV. Therefore, it provides more flexibility and freedom to operate and obviously it's quite popular version. At the same time the disadvantage is complexity is very high due to more number of components.

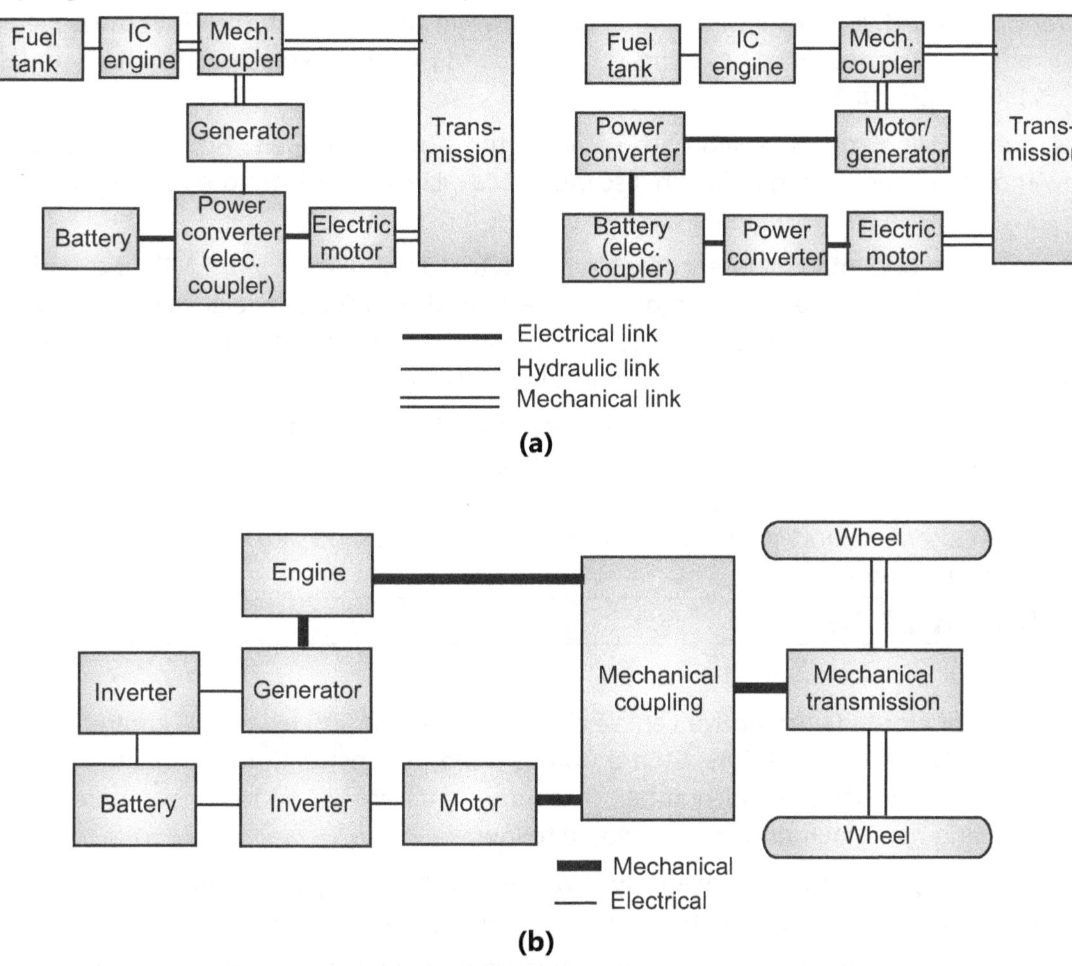

**(a)**

**(b)**

**Fig. 1.8: The Architecture of a Series-Parallel HEV**

**(d) Complex Hybrid:** In this type of HEV multiple motors are used for drive and planetary gears are involved. Generally it's used in 4 wheel drive (4 × 4) or all wheel drive vehicles. The separate drive axles are used as shown in Fig. 1.9. In this system the generator is used to perform series operations as well as it controls engine for it's maximum operating efficiency. Two electric motors are used to get all wheel drive performance and same time it supports for regenerative braking. This also helps for enhancing vehicle stability and antilock braking.

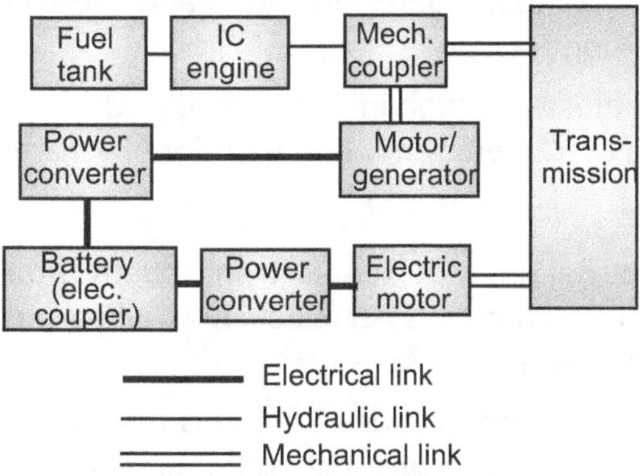

**Fig. 1.9: Complex Hybrid**

The diesel cars are known for fuel efficiency and hybrid diesel cars are much more fuel efficient. The application of hybrid is extended on Trucks and Buses. With unique hybrid conversions delivers better driving patterns and improves substantial fuel efficiency. All types of hybrid concepts are observed on those trucks and buses which provides substantial savings to the owners and becomes viable business proposals. The locomotive engines are also converted as series hybrid and engine plus generator drives trains with hybridization.

Increasing trend of hybrid vehicles in global market is seen in recent trend. For 2000 the developments started and companies like Toyota, Honda, Ford has handsome trends of market share. Nissan, GM, Hyundai and Lexus are other popular OEM in hybrid market.

Toyota Prius, Camry, Highlanders, Avalon are some of the popular models. Honda has Civic, Accord, Insight, CRZ, CRV are few model names. Ford has Fusion, Escape, C-Max, Lincon MKZ are the few models has got good sale in US market.

Increasing trend of hybrid market is encouraging and trends shows SUV or MID size vehicle are more in demand with hybrid cars. Off course the price of all these hybrid vehicles are relatively high and gives benefit of excellent fuel efficiency. Increased components and complexity also add to concerns on safety and reliability. Packaging becomes more complicated and critical which leads to high level of difficulties in service.

Also controls and power management becomes very critical. Some of the challenges in technology need to provide better solutions on Power electronic and electrical mechanism, electromagnetic interference, energy storage systems, regenerative braking controls, Thermal management and overall optimizations are key factors in design and developments of Hybrid vehicles. Therefore, Indian markets are still not open for those high end technology cars in large numbers.

## 1.2 ELECTRIC VEHICLES

### 1.2.1 Introduction

Today's more popular word in automotive vehicles is Electric vehicle (EV) which has created new scenarios in the market mainly due to NON POLUTION Cars. EV are quite simple in construction and relatively easy in operations. The only source of energy is battery and number of components are very low as compare to ICE or HEV. The construction is simple and manufacturing cost is also quite low.

Therefore, customers get products at relatively low price with good performance. The upmost advantage of drastic reduction in overall pollution level in major cities as the population of EV will increase.

The requirement of fuel, complex IC engines and exhaust systems are completely eliminated in EV. Fuel Cell is another source of energy in EV category however it's not considered as cost effective nor efficient in terms of size and weight.

Further The EV are very quiet and it's talked about introduction of noise since it's extremely silent. EV's are fully matured and are able to deliver adequate power and energy with in compact size, weight and cost.

EV is considered as revolution in technology of automotive technology. The vibrations are never noticed in EV as against IC engines; which have multiple rotating, reciprocating and oscillating parts which caused the vibrations. In case of EV motors can create very smooth torque.

Finally the efficiency of motor is almost very high in order to 70% where as IC engine operate at 30 to 37%. The battery and power electronic operates at 90% efficiency and therefore EV's are more sustainable.

### 1.2.2 EV Controlling

The Electric Vehicle consist of electrical drive, power electronic controls for drive, battery, battery management, battery charging system and all other mechanical systems such as wheels, suspension, steering, body, seats and primary safety devices like seat belts and ABS etc.

In the power electronic system high voltage electrical charge is used to drive the vehicle. At the same time low voltage system is required in EV for normal operations such as lighting, horn, tail lamp, motors for windshield wiper wash and window winding mechanism.

Hence the electrical diagrams become complex for Low voltage and High voltage systems. Motor is part of power train which has controller to charge and discharge battery and supply the energy to power train. It mainly controls the speed and torque of motor with the help of power electronics.

The high voltage high current cable length becomes more critical. Therefore, packaging of motor, battery controller plays important role in the system as it's placed in close vicinity the efficiency improves with challenges to handle thermal management.

As a results the cooling system for battery, motor and battery charging controller will get add in the component level and need to fix with better efficient way. This aspect need to taken care for its packaging cost.

The control box receives signal and power, both high voltage for propulsion and low voltage for specific devices i.e. cooling system and then through this control box interface functional channels to power train motor or high voltage battery.

These are functionally separate or merge as per the integrating needs in the system. The control box also receives the signal for velocity of vehicle, throttle position through acceleration pedal. Based on the information received and demand for increases or decreases of speed and torque it makes the decision through microprocessor and send the signal to power train control and adjust the changes in torque and speed request.

Similarly brake pedal gives signal to controlling device through the sensor on brake pedal to apply the brake or release it. Accordingly control box send the signal with required brake force to power train to reduce the speed and torque to motor.

The opportunity of regenerative braking is also captured and send the signal to absorb energy from braking. All these exchange of information is very fast and happen in few micro seconds, therefore decision making and signaling is bit complex and needs good reliable and robust system.

The information exchange is normally called as Control area network (CAN) bus. This is the communication signals in multiplex levels. Some sort of protocol has been observed when multiple signals are processed together and passed through and shared by common medium.

In other words, some sort of priority based signals flow when trying to share the same physical medium. When there are some of the signals of low priority such as door lock switch to turn on are kind of afford to wait.

Whereas braking or turn steering are very important safety signals which are priority based, these need to transmitted immediately. These are some of new protocols that allows by CAN and activate the functions.

Safety Critical functions are identified and microprocessor allows it on priority. The additional hardware based back systems and communication mechanisms are planned to avoid any failures.

## 1.2.3 EV Functionality

EV are extended versions of IC engine functional loads and many common applications are seen in EV. The regular vehicle level loads are electrical devises and other mechanical devices.

Low voltage 12V system loads are very common with ICE vehicles and known as auxiliary load. These are safe in terms of risk level. The many of such low voltage components from existing ICV are advantageous rather than transforming to high voltage parts. Following auxiliary load components are used in EV without conversion.

Brake motor Air conditioning motor, radiator fan, various pumps on windshield washer, lubrication systems, motors for window winder, door lock, wiper, outside rear view mirrors, various lights tail lamp, side indicator, brake indicator emergency indicator, turn signal and Radio Music system, GPS, TV screen all are considered as non-motor and low voltage systems. Also engine controllers, transmission controller vehicle body controller, various computational microprocessors, digital signal processors are non-motor low voltage electrical systems are used inline with ICE systems.

The various pumps and fans produces 100~200W or less, and some of the motors and lights are even less than 50 or100W. These loads are fed by 12V battery and battery is charged with small onboard chargers or generator.

The High voltage load (i.e. rage from 48V to 600V) the propulsion system and motors are considered in EV. The load from propulsion systems are regenerative braking which runs from 50kW to 500kW in EV and HEV.

Sometime low voltage charging system can be part of main high voltage system and drawn from DC-DC converter. Thus the multiple architecture is possible for combine high and low voltage systems.

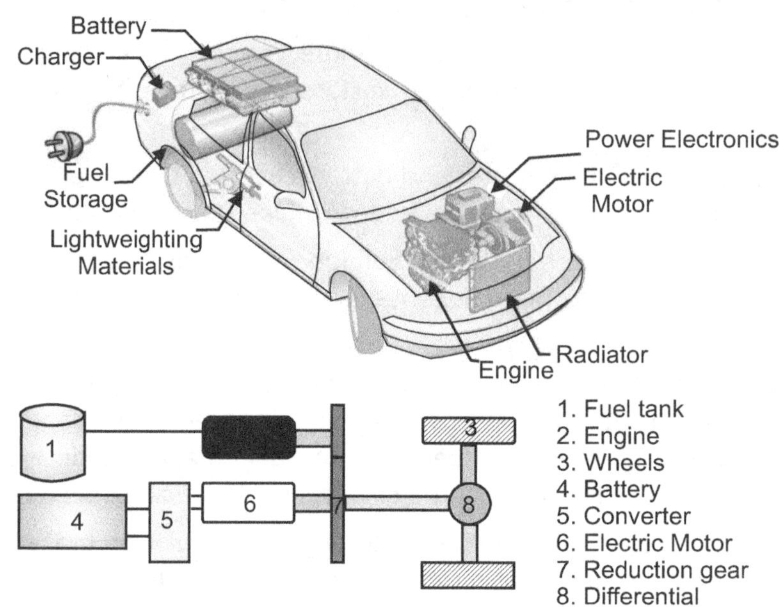


**Fig. 1.10: Electric Vehicle**

## 1.2.4 Electrical Energy Storage Device: Battery

Battery is well known as a electrical charge storage device. Similarly, Super capacitors & Ultra capacitors also store the electrical energy. These storage devices are charged from ICE through generator or directly from grid power and then it can supply power for propulsion of EV and HEV. This storage devises help in recuperating regenerative braking energy and stores it for future usage.

The basic element of battery is several cells together constitute modules and several modules constitute battery pack. These cells are modules are connected in series and parallel and complete battery pack from the range of 12 V to 240 V are normally configured. (Refer Fig. 1.11)

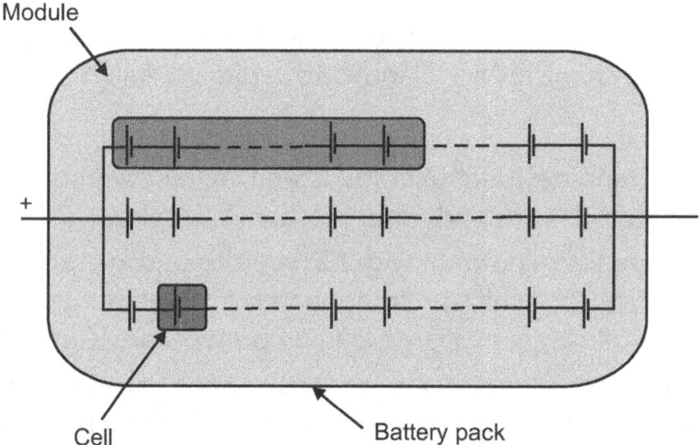


**Fig. 1.11: Relationship Between Cell, Module and Battery Pack**

The charge and discharge is controlled by Battery management system which is essential in EV and HEV purpose. A battery is an energy intensive device and for given size and weight it can store much more energy. In case of Ultra capacitors is power intensive storage and relatively low energy is stored as compare to battery.

It attributes to low internal resistance and faster responsive as compare to battery. In terms of consistency in energy supply battery is highly non linear electrochemical device, also it is very sensitive about rise in temperature in terms of charge and discharge.

Therefore, w/o controller device it's can't be efficient. Generally, battery has limited few thousands of charge and discharge cycles therefore life is limited period.

Characterization of battery: Battery capacity implies the amount of electrical charge, that it can supply before it's fully discharged. It's depend on temperature of battery and level of discharge. The voltage lower to certain specified value that it can't further charge to a load.

The SI unit of battery charge is Coulomb (C) which is ampere-second or ampere- hour (Ah). A battery of 100Ah indicates 1Amp current can be available for 100 hrs. as current goes up time will get dived and performance duration will reduce.

While charging the amount of charge that battery can accept and can release the charge during discharge mode is also important.

Thus efficiency parameters of battery are dependent on charging and discharging current rate. This is known as C-rate and defined as charging current which will charge a complexly discharged battery to it's rated ampere-hour (Ah) during charging.

Similarly, discharged fully charged battery (at rated Ah charged) to it's fully discharged condition. Crate is equal to rated Ah but unit is Ampere.

The stored energy in battery depends on terminal voltage and amount of charge stored within it at particular temperature. Energy is charge (Ah) multiply by voltage and unit of energy is watt–hr = E = amp-hr × v.

The state of charge (SOC) is an important parameter related to battery which is quite often used. The SOC measured as Amp-hr balancing or EMF based as direct method. For more accurate perspective it's desirable to use both the methods in combination and alternately depending the state of battery. Battery current and voltage is measured across the terminals.

Temperature is not the characteristic to be measured on battery but ambience and operating temperature is considered over the measurements. Therefore, battery impedance is considered which is function of V, I and T. Only control is possible for voltage or current but not independently.

Some time while charging outside temperature is controlled and can get better charge to battery thorough the charger or generator or Fuel cell sources. External fan or cooling devices are deployed for such requirements.

The status of battery is measured in open circuit voltage and use a lookup table to correlate with SOC. Generally, manufactures data table is made available. For impedance (Z) direct calculation or through matching with model involving resistors and capacitors are considered. The other method of apply current steps and monitor the voltage time constant r after the application of step. The real time environment; all these can be implemented when system is delivering load.

One way is to apply just very small percentage of the total load current through a current command to battery and monitor the voltage charge and time contestant and record these.

It will not be very significant impact to main load however provide real-time information.  In addition to it we have to monitor temperature T since SOC is dependent on it.

The end result is expression of SOC = f (V, I, T, Z, and r). where as

| | | |
|---|---|---|
| where, | V = | Battery open circuit voltage. |
| | I = | Battery current. |
| | T = | Battery temperature. |
| | Z = | Battery Impedance. |
| | r = | Battery voltage relaxation time after application of step current. |

This is bit simplified version which may be adequate for power management, but in general it could be highly nonlinear which may be not be expressed as direct functional from.

Specific power is the rated power of battery for a given weight and there is volume related specific power defined by rated power of the battery for given size or volume of battery. i.e. watt per Kg or watt per cubic meter.

Similarly, specific energy watt-hr per kilogram and specific energy watt-hr per cubic meter is taken into consideration. The relationship between specific energy and specific power is represented by using graph is plotted with x-axis to specific power and y axis specific energy. This is useful for compare various batteries or type of energy devices (ultra capacitor) refer the Table 1.1.

**Table 1.1**

| Energy source | Specific energy (watt-hr/kg) |
| --- | --- |
| Gasoline | 12500 |
| Natural gas | 9350 |
| Methanol | 6200 |
| Hydrogen | 28000 |
| Coal | 8200 |
| Lead acid battery | 35 |
| Nickel metal hydride battery | 50 |
| Lithium –polymer battery | 200 |
| Lithium-ion battery | 120 |
| Sodium sulfur battery | 150-300 |
| Flywheel (steel) | 30 |
| Ultra capacitor | 3.3 |

## 1.2.5 Charging and Discharging Efficiency of Battery

Ampere-hour efficiency is the ratio between discharge electric charge given out during discharging a battery and the electric charge needed to battery to returned to previous charge level. The quantity are not equal and could be in order of 65-90% is is generally efficiency of battery. The battery chemistry, operating temperature and rate of charge all these factors are important for efficiency.

The other aspect of variation of equivalent internal resistance and chemical process during charging and discharging affects the efficiency of battery. One more factor of time duration of charging or discharging is done slow (flow of current) or quick (high current).

EV and HEV batteries can undergo few hundreds to couple of thousands (i.e. 300~5000) deep charges is known as battery life. DOD is depth of discharge depends upon battery chemistry and usage pattern. In general, life is defined for minimum 600 deep cycles for EV batteries.

Fig. 1.12 shows % life reduction with temperature.

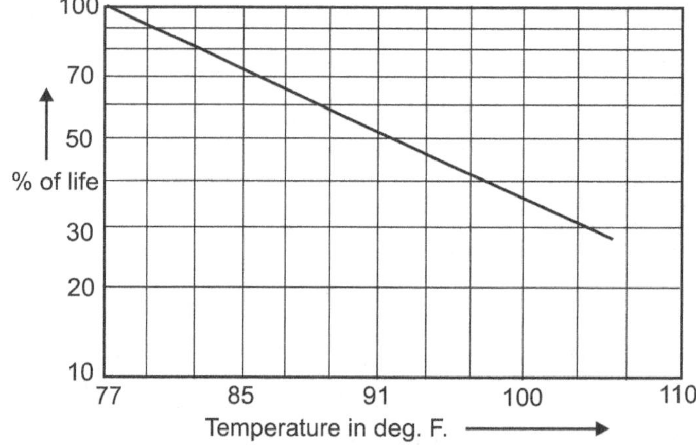


**Fig. 1.12: Battery Percentage Life Reduction with Temperature**

From user's perspective goal is to get maximum life and energy capacity out of battery. Following good practices are adopted while charging of battery –

1.  **Lead acids battery during charging:** It is to be charged immediately it's usage and full discharge is to be avoided. Overcharging and overheating is to be avoided. Constant voltage charging is followed by floating condition should be applied.

2.  **Lead Acid battery during discharge:** Avoid full cycle discharge, 80% depth of discharge is ok, recharge more often.

3.  **Lithium Ion during charging:** Charge often, full discharge to be avoided over-cycling of charge–discharge shall be avoided, constant voltage charging should not be followed by any trickle charge, fast charge possible, avoid overheating.

4.  **Lithium Ion during discharging:** similar to lead acid discharging practices.

5.  **NiMH during charging:** Run it full down once in every three month of usage. Avoid overheating, fast charging ok but never go for slow charging, Constant current charge when it's full should be applied.

6.  **NiMH during discharging:** Avoid too many full cycle of charge and discharge. Use it 80% depth of discharge.

**Battery Management System (BMS):** The main function of battery management is to control the temperature during charge and discharge and control environmental conditions ambient temperature and humidity around the battery system and controller.

Battery management implies that the parameters external to battery are to be influenced according to the different battery chemistries through some control mechanism. This includes data acquisition of temperature, current, voltage, humidity etc.

The designed and actual measured data is compared and controlled. The thermal management and current management are the constantly under control. Power electronic and matrix switching many modules during the control. All sensors are required to be calibrated and artificial intelligence techniques are utilized by using voltage sensors and derived current based on AI algorithms. The Fig. 1.13 shows the BMS architectures.

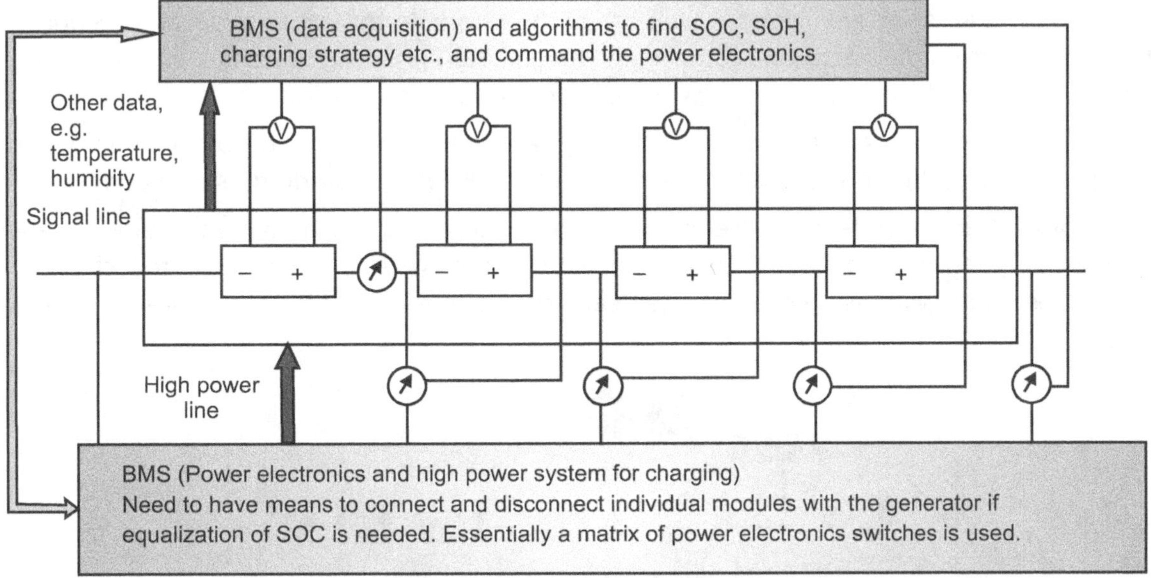


**Fig. 1.13: Battery Management System Level Architecture**

The BMS architecture shows the power electronic which deals with high current and provides interface between battery and load, generators. It should be possible to implement architectural reconfiguration of battery modules such that the generator can selectively charge one or more modules, independent of other modules. Similarly, it should allow discharge selectively if it's necessary. This may require provision of bleed resistor in parallel with battery which could dissipate the energy of battery for discharge.

The comparison various battery types & energy storage for HEV.

**Table 1.2**

| Storage Technology | No. of cycles | Efficiency | Specific power | Specific energy |
|---|---|---|---|---|
| | Life | % | watts/kg | watt-hrs/kr |
| Lead Acid | 500 - 800 | 50 - 92 | 180 | 30 - 40 |
| Li-ion polymer | 500 - 1000 | 80 - 90 | > 3000 | 130 - 200 |
| NiMH | 500 - 1000 | 66 | 250 - 1000 | 30 - 80 |
| Ultra capacitor | 10,00,000 | 90 | 1000 - 9000 | 0.5 - 30 |

Usage patterns of different types of batteries :

**Lead Acid:** Automobile Industry for starting of engine, Domestic purpose for current supply in case of grid power supply fails or not available. Generally, rage from 6-24 Volts are commonly used.

**Li-ion** : EV and HEV for power supply and in cell phone and electronic devices for small range power application.

**Nickel Metal Hybrid Battery:** Hybrid vehicles for high voltage applications and Aircraft electronic instruments.

**Ultracapacitor:** It is quite different from batteries energy storage within it and do not change to chemically during charge and discharge. Moreover, it's used rapid generations of power in regenerative braking of HEV, in electronic circuits and appliance of high end products.

**Conclusion:** *The battery and battery management are the crucks of the technology to be deployed on various EV, HEV and advance vehicles. It's applications and usage will be the increasing trend for next decade. The EV are the presence and Hybrid will be future of Indian mobility.*

## 1.3 SAFETY IN AUTOMOBILES

### 1.3.1 Introduction

Work in area of passive safety in vehicles is well established domain for last four decades. Significant work has been done by many vehicle manufacturers, component manufacturers and universities in abroad. Thanks to their co-ordinated efforts due to which vehicles are now much safer to drive on roads.

Moving forward, active safety and autonomous driving has evolved in the automotive landscape in last decade or so. These are promising technologies to enhance occupant safety while providing driving comfort.

The overview of each of these technologies and considers the possibility of synergy between them. The close coordination between passive safety, active safety and autonomous systems can help to achieve best possible scenario to deploy restraints systems such as airbags, seat belt pre-tensioners along with pre-crash braking.

**Passive Safety:**

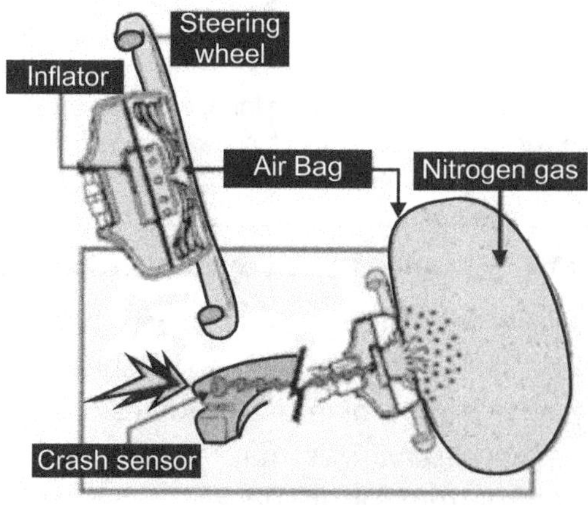


**Fig. 1.14: Airbag System**

The main function of passive safety features in a car is to keep the driver and passengers protected within the vehicle from various crash forces. Airbags, seat belts are restraints systems designed to absorb crash forces and minimize hard contact of occupants with vehicle interiors.

The vehicle are designed to absorb the impact load during the crash or hit of two moving cars or stationary objects and protect the passagers inside the car as well outside pedestrian on the road. The body structure are made such as panel thickness, weld quality, geometry of parts and choice of materials with it's strength are some of the factors considered during design of vehicles. Additionally, dashboards, seats, seat belts and steering column are designed to soften the hard impact to occupants in high speed crashes. Now-a-days most of the vehicles comply with Indian norms. Introduction of passive safety in India has been rather late.

However, with Indian regulations push and Indian automotive industry to sustain with global competition and with awareness among people, the safety regulation in India has been introduced from Oct. 2017.

## 1.3.2 Active Safety

Active safety systems detect the obstacles of various size and nature. Using this information collision can be avoided or speed of contact with the obstacle is reduced considerably. The sensor and camera based CAN communications and microprocessors ECU device help to regulate speed of vehicle. The decision of brake the speed is done by intelligent controls.

Various active safety systems are available such as driver assist, Advanced emergency braking system (AEB), Anti-lock braking system, Electronic Stability control (ESP) to name a few.

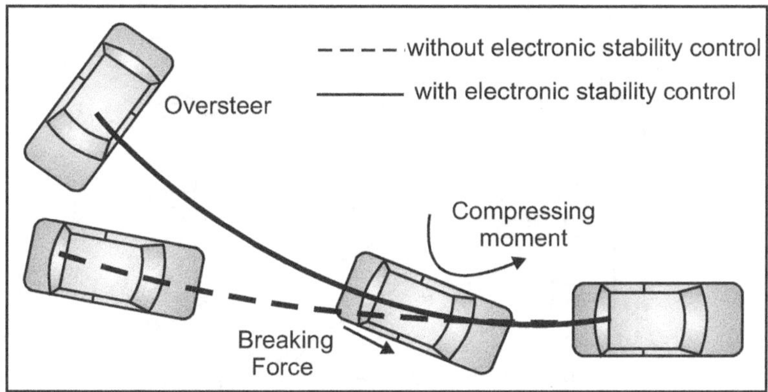


**Fig. 1.15: Electronic Stability Program**

**(a) Electronic Stability Program**: The system controls the longitudinal and lateral behavior of the vehicle.

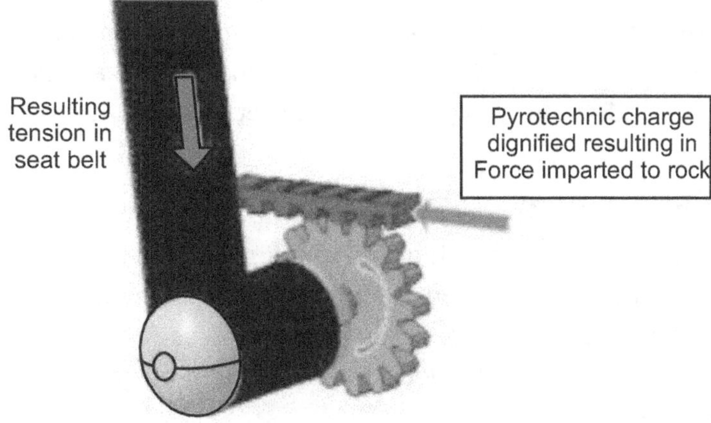


**Fig. 1.16 (a)**

While longitudinal behavior is controlled through acceleration and braking, the lateral behavior controls the roller and lateral skidding.

The lateral velocity and skidding angle are important parameters calculated by ESP and can be used in conjunction with side acceleration and pressure sensors to best estimate the side airbag fire times.

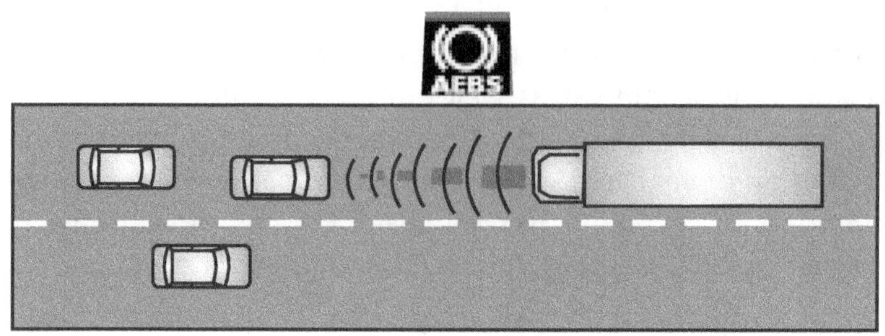


**Fig. 1.16 (b) : Advanced Emergency Braking System**

**(b) Advanced Emergency Braking System:** The system monitors the vehicles in front and evaluates relative speed and gap required and applies brakes to keep the proper distance from the vehicle ahead. AEBS allows to keep safe distance (i.e. 20 ft) between two cars. The leaser signals transmitted and received back to system and which governs the safe distance. In case of any emergency if next car stops the following car brakes without action form driver by AEBS and reduce the speed for safety.

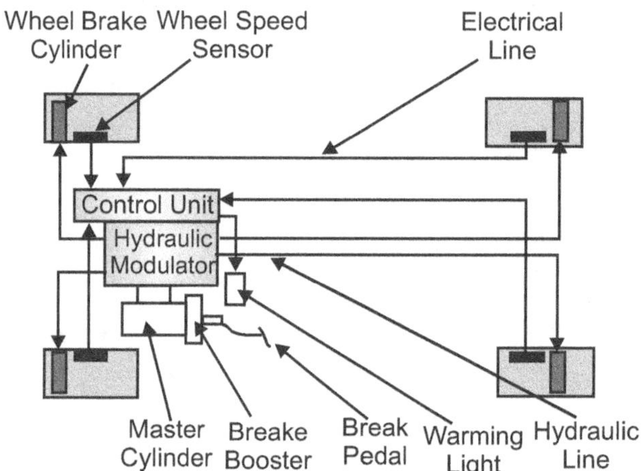


**Fig. 1.16 (c): Anti-skid Braking System**

**(c) Anti-lock Braking System:** Anti-Lock braking system mainly known as ABS. Basically, it allows the wheels to maintain traction control with road surface and prevents the wheels from locking up and avoid uncontrolled skidding. Now-a-days most of the cars come with antilock-braking system as standard. During the panic braking the vehicle is steered by driver and taken away from the target which may hit. This ABS feature safeguard the passengers and avoid collegian.

## 1.3.3 Autonomous Driving Systems

The autonomous driving uses complex sensing and processing system to analyze the vehicle internal and external environment. The benefits are extensive, such as to free human drivers from fatigue, to ease road congestion and to lower fuel consumption. While existing technologies can assure high fidelity sensing and robust control, the challenges lie in the interactions between the automated vehicle and other road participants such as manually driven vehicles and pedestrians.

The autonomous vehicle hierarchical architecture is shown alongside. The system uses sensing, decision making and control strategies to accurately map the external environment and execute essential controls such as steering, braking and acceleration.

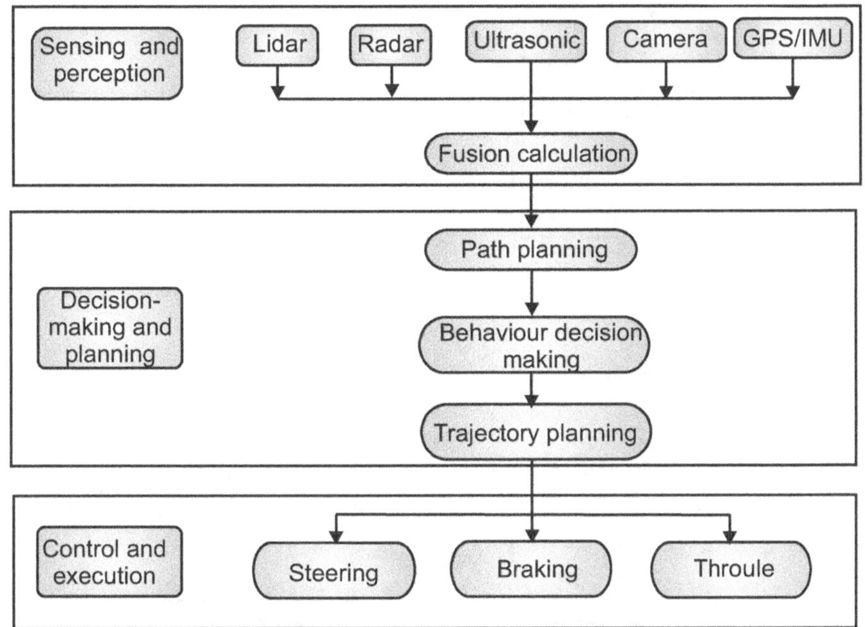


**Fig. 1.17: Autonomous Vehicle Architecture**

**Integrated Approach:** Although these technologies do wonderful job in achieving complex driving tasks, they still do it independently.

If the algorithms developed to execute such independent tasks are brought together and made to work in synergy, new set of functions evolve. Some of the scenarios where these systems can work together is explained.

**Case 1**: The adaptive cruise control which is a part of Autonomous driving system can continuously track the external environment such as vehicle in front, on the side and rear. Also the condition of roads can be tracked using different sensing technologies.

If the vehicle in front were to lose the control and over steer, the adaptive cruise control can send the signal to ESP ECU and Restraint ECU via CAN network.

Depending upon the closing distance, the Restraint ECU can prepare for imminent crash and deploy pre-crash reversible restraints systems such as reversible belt tensioner.

This function fixes the occupant in the seat and therefore helps improve the occupant's position before crash and avoid the occupant contact with deploying airbag.

Further the Advanced emergency braking system can apply brakes thus reducing the speed of impact. This drastically reduces the injury.

**Case 2:** Usually in pole crashes where vehicle skids off the road and hits the pole, tree or a barrier, the occupant moves closer to the door.

The decision to deploy the airbag can be effectively taken if the signals such as yaw rate, lateral velocity and slip angle are available.

Using these signals ESP ECU can send critical threshold signal to Restraint ECU to prepare for imminent crash.

**Case 3:** The pedestrian is detected by the Sensing systems mounted in front of the vehicle. The Vehicle intelligent system can send the signal via CAN network to all the participating ECUs.

This leads to breaking of the vehicle and avoid the contact with pedestrian. However if the vehicle speed is too high to avoid the impact, then along with braking function, Restraint ECU can get ready to activate the active bonnet in the vehicle.

The data processing that integrates assistance and safety functions is shown alongside.

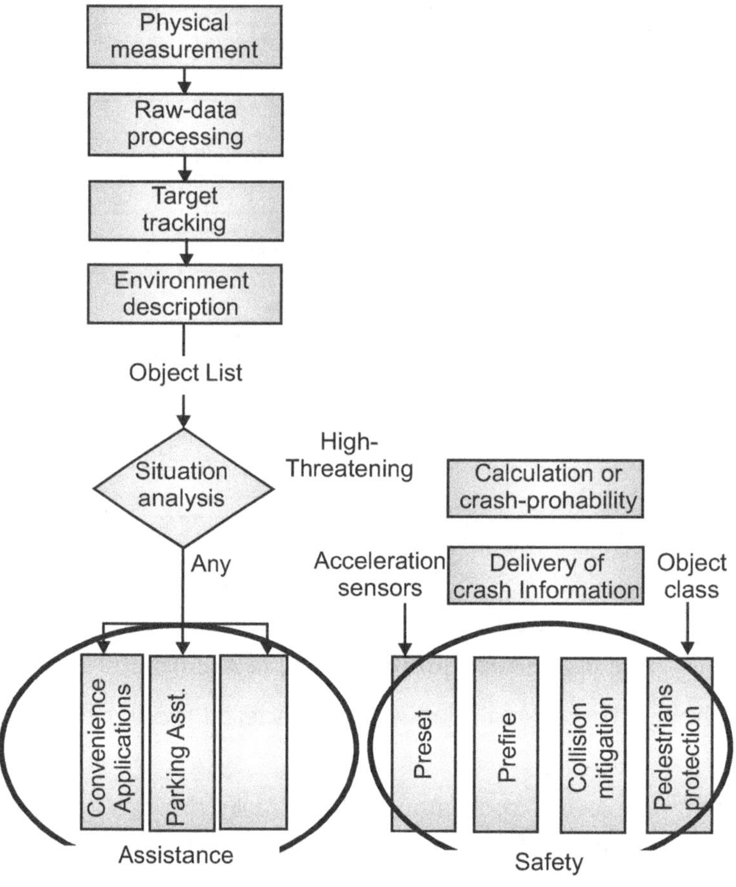


**Fig. 1.18: Algorithm of Safety Devices**

**Conclusion:** Integrated approach is way forward to mitigate/eliminate the occupant injury during crash. However the complex algorithm, data processing hardware and sensing technologies are expensive and needs to be evaluated on cost vs. benefit ratio.

## *Multiple Choice Questions*

1. A Type-1 hybrid vehicle will enter READY TO DRIVE (KOER) mode but will not crank the internal combustion engine (ICE). When the vehicle is pushed in NEUTRAL gear range, the engine spins. Which of these could be the cause?

   (a)  A depleted high voltage battery.

   (B)  A seized MG2.

   (c)  A failed crankshaft position sensor.

   (d)  A seized plenetary gear set.

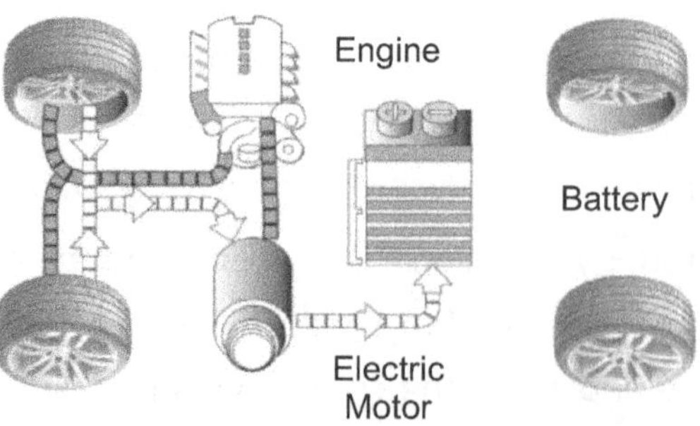


**Fig. 1.19**

2. During deceleration, the arrows on the power flow display appear as shown. This indicates ......
   (a) The engine is running with energy flowing to the drive wheels.
   (b) The engine is running and driving the motor/generator to recharge the battery pack.
   (c) Energy from the battery pack and the engine are blended and flowing to the drive wheels.
   (d) Energy from the drive wheels is being used to recharge the battery pack.

3. A Type-2 hybrid vehicle with a belt-type continuously variable transmission (CVT) bucks and jerks when accelerating from a stop.

   Technician A says that using the wrong type of transmission fluid could be the cause.

   Technician B says that a failing split pulley/steel belt could be the cause. Who is right?
   (a) (a) only Technician A                    (b) (b) only Technician B
   (c) Both Technician A and B (a) and (b)       (d) Neither of Technician A or B (a) nor (b)

4. Plugin hybrids running on electricity powered from coal can significantly reduce greenhouse gases.
   (a) True
   (b) False

5. Which is a valid issue with electric vehicle batteries?
   (a) They need to be replaced every 5 years.      (b) They cannot be recycled.
   (c) They are expensive.                           (d) They are less safe than gasoline engines.

6. Nissan and Toyota are currently testing electric cars that store electricity from off-peak hours and supply it to households during peak hours or emergencies. The vehicle to home (V2H) technology is expected to appear in Japanese markets in 2012 ?
   (a) True
   (b) False

7. Everyone charging their electric car at the same time will cause brownouts and blackouts.
   (a) True
   (b) False

8. The consideration involved in the selection of the type of electric drive for a particular application depends upon ......
   (a) Speed control range and its nature          (b) Starting Nature
   (c) Environmental condition                      (d) All of the above

9. The capacity of a battery is expressed in terms of ......
   (a) Current rating      (b) Voltage rating      (c) Ampere hour rating   (d) None of the above

10. The storage battery generally used in electric power station is ......
    (a) Nickel-cadmium battery                      (b) Zinc carbon battery
    (c) Lead-acid battery                            (d) None of the above

11. Battery charging equipment is generally installed ......
    (a) In well ventilated location.
    (b) In clean and dry place.
    (c) As near as practical to the battery being charged.
    (d) In location having all above features.

12. Batteries are charged by ......

    (a) Rectifiers.	(b) Engine generator sets.

    (c) Motor generator sets.	(d) Any of the above.

13. Battery container should be acid resistance therefore it is made up of ......

    (a) Glass	(b) Plastic	(c) Wood	(d) All of the above

14. Local action in a battery is indicated by ......

    (a) Excessive gassing under load conditions.

    (b) Excessive drop in the specific gravity of electrolyte even when the sale is on open circuit.

    (c) Both (a) and (b).

    (d) None of the above.

15. Which of the following battery is used for aircraft?

    (a) Lead acid battery.	(b) Nickel-iron battery.

    (c) Dry cell battery.	(d) Silver oxide battery.

16. When two batteries are connected in parallel, it should be ensured that ......

    (a) They have same emf.	(b) They have same make.

    (c) They have same ampere hour capacity.	(d) They have identical internal resistance.

17. A floating battery is one ......

    (a) Which gets charged and discharged simultaneously?

    (b) Which supplies current intermittently and also during off cycle gets charged?

    (c) In which battery voltage is equal to charger voltage?

    (d) In which the current in the circuit is fully supplied by the battery?

18. What's the critical part of building an electric car?

    (a) Body	(b) Wheel	(c) Battery	(d) Engine

19. There is no gasoline engine in an electric car, so what runs that thing?

    (a) Internal combustion engine	(b) Hamsters

    (c) Electric motor	(d) None of these

20. Electric cars emit no tailpipe pollutants.

    (a) True

    (b) False

**Answers**

| Question No. | 1 | 2 | 3 | 4 | 5 | 6 | 7 | 8 | 9 | 10 |
|---|---|---|---|---|---|---|---|---|---|---|
| **Answer** | (d) | (d) | (c) | (b) | (c) | (a) | (b) | (d) | (c) | (c) |
| **Question No.** | 11 | 12 | 13 | 14 | 15 | 16 | 17 | 18 | 19 | 20 |
| **Answer** | (d) | (d) | (d) | (d) | (b) | (a) | (b) | (c) | (b) | (a) |

# Chapter 2

# PROCESS ENGINEERING

**Weightage of Marks = 10, Teaching Hours = 06**

### Syllabus

2.1.  Process Boiler, Steam, Condensate loop in process industry, Steam Distribution System, Steam Traps, Function of Steam Traps and Type of Steam Traps.

2.2.  Introduction of Ultra Supper critical Boiler, Principle, Working, advantages, application.

2.3.  Hyperbolic Cooling Towers.

2.4.  Waste heat recovery in process industries

### About this Chapter

At the end of this chapter, students will be able to:

- Apply Heat Engineering in Process Boiler
- Understand Ultra Supper Critical Boiler (USC)
- Understand Hyperbolic Cooling Towers
- Waste heat recovery system used in process industries

## 2.1 INTRODUCTION

**Steam:**

Steam is a vapour form of water. Steam has been a popular mode of conveying energy since the industrial revolution. Steam is used for generating power and heating in process industries such as sugar, paper, fertilizer, refineries, petrochemicals, chemical, food, synthetic fiber and textiles.

**Boiler:**

It is a closed vessels in which water gets converted into steam with the addition of heat energy which is generated by combusting fuel (may be solid, liquid and gaseous) in combustion chamber or furnace.

**Process Boiler:**

The boilers used in process industries for heating purpose are called as process boilers. The pressure and temperature of the steam generated in these boilers is basically low. Process boilers are capable of generating saturated steam at a pressure of 10 to 15 bar. The saturated steam is then led to the point of application and the latent heat rejected during the condensation of the saturated steam is used for process heating.

**Sensible Heat of water:**

The amount of heat required by 1Kg of water to raise its temperature from freezing temperature up to the boiling temperature or saturation temperature is called sensible heat

Sensible heat is called as Liquid enthalpy which is the "Enthalpy" (heat energy) in the water when it has been raised to its boiling point to produce steam, and is measured in kJ/kg, its symbol is hf.

$$Q = m \times C_{pw} \times (T_w - T_i)$$

where, Q = Heat, m = Mass of substance,

$C_{pw}$ = Specific heat of constant pressure of water

$T_w$ = Temperature of water after heating,

$T_i$ = Initial temperature

## Latent Heat:

The amount of heat required by water which is at boiling or saturation temperature to convert it into vapor form or steam at constant temperature such a heat is called latent heat. When the latent heat is acquired by the water only phase change will take place but there is no change in temperature.

Latent heat is also called **Enthalpy of Evaporation.** When the steam condenses back into water, it gives up its enthalpy of evaporation, which it had acquired on changing from water to steam. The enthalpy of evaporation is measured in kJ/kg.

Its symbol is $h_{fg}$. In process industries the latent heat given out by condensing saturated steam are used for heating applications.

## Saturation Temperature:

The temperature at which water boils, also called as boiling point or **saturation temperature** increases as the pressure increases. When water under pressure is heated its saturation temperature rises above 100°C and the usable heat energy in the steam (enthalpy of evaporation) decreases.

## The Steam Phase Diagram

The relationship between the enthalpy and the saturation temperatures at various different pressures is known as a phase diagram.

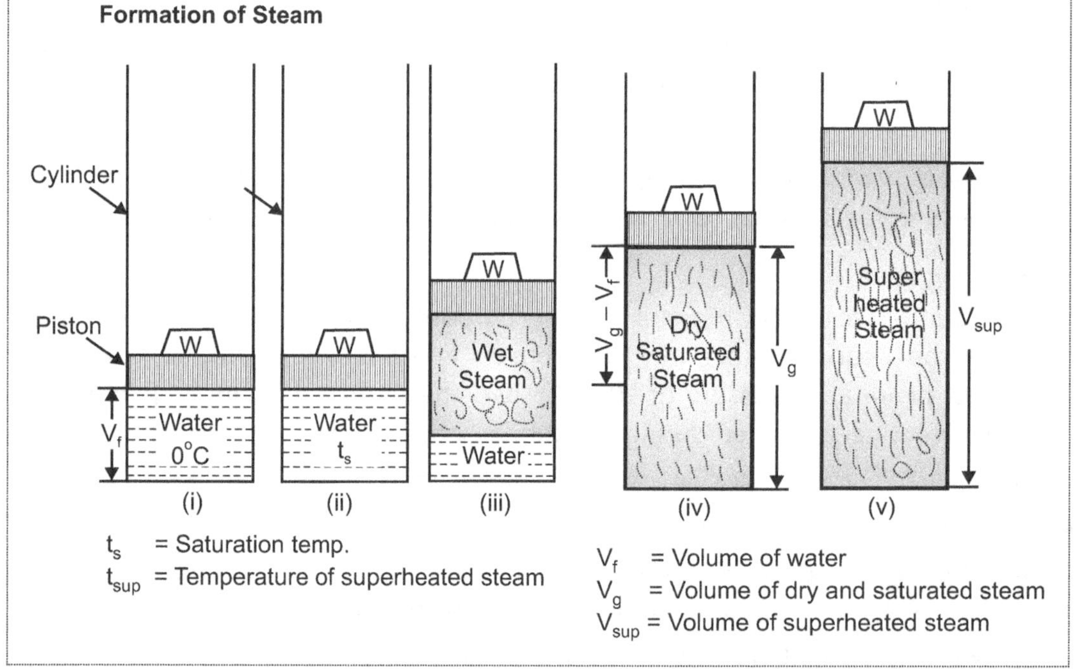


**Fig. 2.1: Formation of Steam**

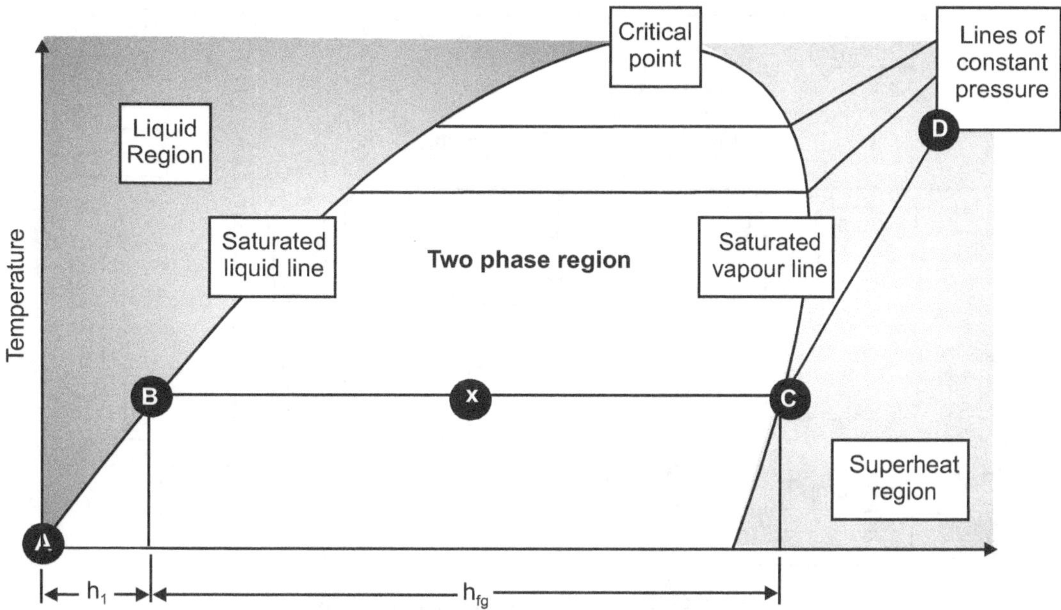


**Fig. 2.2: Steam Phase Diagram**

As water is heated from 0°C to its saturation temperature, its condition follows the saturated liquid line until it has received all of its liquid enthalpy, $h_f$, (A - B) this phase of heat is call sensible heat.

Continuously adding heat the temperature remains constant, where the phase changes to saturated steam and there continues increase in enthalpy, $h_{fg}$, (B - C).

The condition moves from the saturated liquid line to the saturated vapour line, as the steam/water mixture increases in dryness, the point is mid-point of B - C that means the dryness fraction is 0.5 (50%), when the point moves toward C side the dryness increases and when it move towards B side the dryness decreases. When point reaches to saturated vapour line the steam attains 100% dryness.

When it reaches the saturated vapour line, the steam receives all of its enthalpy of evaporation. On continuously heating after this point, the temperature of the steam will begin to rise as superheat (C-D).

The region to the left of the saturated liquid line only water exists, and in the region to the right of the saturated vapour line only superheated steam exists and the region in between or enclose by saturated liquid and saturated vapour lines in which a steam/water mixture exists - wet steam.

The point at which the saturated liquid and saturated vapour lines meet is known as the critical point. As the pressure increases towards the critical point the enthalpy of evaporation decreases, until it becomes zero at the critical point. This suggests that water changes directly into saturated steam at the critical point.

Above the critical point only gas may exist. The gaseous state is the most diffuse state in which the molecules have an almost unrestricted motion, and the volume increases without limit as the pressure is reduced.

**Condensate loop in process industry:**

**Steam distribution system**

The steam distribution system is the essential link between the steam generator and the steam user. Whatever the source, an efficient steam distribution system is essential if steam of the right quality and pressure is to be supplied, in the right quantity, to the steam using equipment. Installation and maintenance of the steam system are important issues, and must be considered at the design stage.

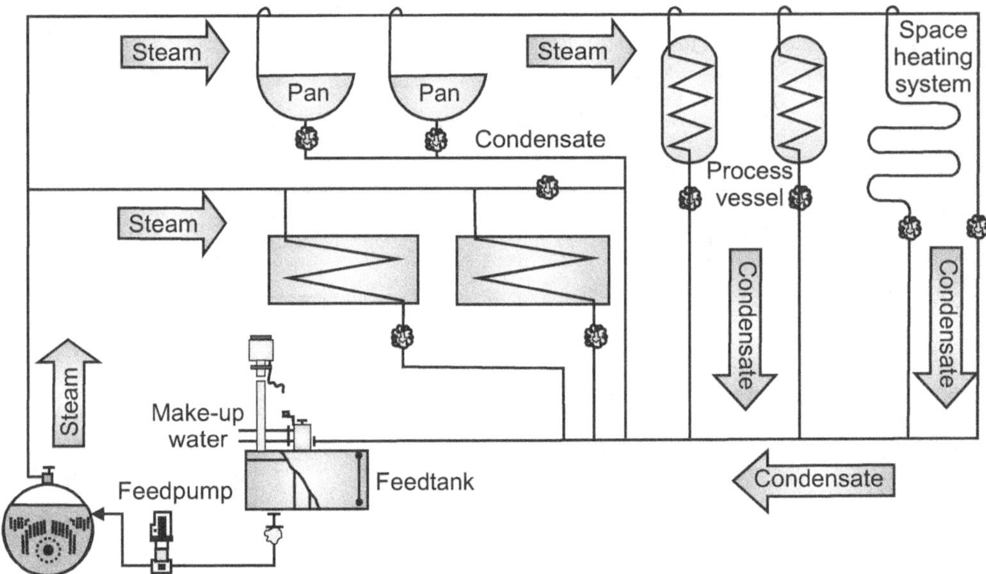


**Fig. 2.3: Steam Distribution System**

As steam condenses in a process, flow is induced in the supply pipe. Condensate has a very small volume compared to the steam, and this causes a pressure drop, which causes the steam to flow through the pipes. The steam generated in the boiler must be conveyed through pipe work to the point where its heat energy is required. Initially there will be one or more main pipes, or 'steam mains', which carry steam from the boiler in the general direction of the steam using plant. Smaller branch pipes can then carry the steam to the individual pieces of equipment. A typical steam distribution system is shown in Fig. 2.3. It must be noted that pressure reducing valves (PRV) are used to reduce the pressure of steam as per the requirement of process device.

The distribution pressure of steam is influenced by a number of factors, but is limited by:

- The maximum safe working pressure of the boiler.
- The minimum pressure required at the plant.

As steam passes through the distribution pipework, it will lose pressure due to:

- Frictional resistance within the pipework.
- Condensation within the pipework as heat is transferred to the environment.

### Steam Traps

The purpose of installing the steam traps is to obtain fast heating of the product and equipment by keeping the steam lines and equipment free of condensate, air and non-condensable gases. A steam trap is a valve device that discharges condensate and air from the line or piece of equipment without discharging the steam.

### Functions of Steam Traps

The three important functions of steam traps are:

- To discharge condensate as soon as it is formed.
- Not to allow steam to escape.
- To be capable of discharging air and other incondensable gases.

### Types of Steam Traps

There are three basic types of steam trap into which all variations fall, all three are classified by International Standard ISO6704:1982.

### Thermostatic (operated by changes in fluid temperature)

The temperature of saturated steam is determined by its pressure. In the steam space, steam gives up its enthalpy of evaporation (heat), producing condensate at steam temperature. As a result of any further heat loss, the temperature of the condensate will fall. A thermostatic trap will pass condensate when this lower temperature is sensed. As steam reaches the trap, the temperature increases and the trap closes.

### Mechanical (operated by changes in fluid density)

This range of steam traps operates by sensing the difference in density between steam and condensate. These steam traps include' ball float traps' and 'inverted bucket traps'. In the 'ball float trap', the ball rises in the presence of condensate, opening a valve which passes the denser condensate. With the 'inverted bucket trap', the inverted bucket floats when steam reaches the trap and rises to shut the valve. Both are essentially 'mechanical' in their method of operation.

### Thermodynamic (operated by changes in fluid dynamics)

Thermodynamic steam traps rely partly on the formation of flash steam from condensate. This group includes 'thermodynamic', 'disc', 'impulse' and 'labyrinth' steam traps.

### Condensate Discharge

Inverted bucket and thermodynamic disc traps should have intermittent condensate discharge. Float and thermostatic traps should have a continuous condensate discharge. Thermostatic traps can have either continuous or intermittent discharge depending upon the load. If inverted bucket traps are used for extremely small load, it will have a continuous condensate discharge.

### Flash Steam

Flash steam is a name given to the steam formed from hot condensate when the pressure is reduced.

Flash steam is produced when high pressure condensate is discharged to a lower pressure.

This shall not be mistaken for a steam leak through the trap. The users sometimes get confused between a flash steam and leaking steam. The flash steam and the leaking steam can be approximately identified as follows:

- If steam blows out continuously in a blue stream, it is a leaking steam.
- If a steam floats out intermittently in a whitish cloud, it is a flash steam.

## 2.2 INTRODUCTION TO ULTRA SUPERCRITICAL BOILERS

Depending upon the steam pressure boilers are classified as sub-critical, supercritical and ultra-supercritical boilers. Boilers operating with steam pressure below 221 bars are called sub-critical boilers. Supercritical boilers have pressure above 221 bar and superheated temperature upto 593°C. whereas steam pressure about 225 bar and superheated temperature more than 593°C are usually called ultra-supercritical boiler. The ultra-supercritical boiler have maximum temperature and pressure range among all boilers. The power plant efficiency in the range of 38% to 42% for sub-critical, 42% to 46% in supercritical boiler and 46% to 48% in an ultra-supercritical boiler. It is possible to use high temperature steam up 620o C due to the development of chromium and nickel based super alloy material. The world's first supercritical boiler Philo power plant of USA 125 MW was commissioned as early as 1957. Large supercritical unit of 1000 MW capacity are in operation in Japan and USA since 1970.

Most of the supercritical and ultra-supercritical boiler are operating on Pulverized coal (PC) technology some boiler also available on circulating fluidized bed combustion(CFBC) technology world largest and first once through supercritical circulating fluidized bed combustion (CFBC) boiler of 460 MW capacity designed and executed by forster wheeler in USA Since 2009.ultra-supercritical boiler are in operation in Japan Germany Denmark and China number of 1000MW ultra-supercritical boiler with steam parameter of 260 bar 600°C pressure and temperature with tendem compounding 4 casing, 4f low type turbine  are operating in China.

**Benefit of supercritical and ultra-supercritical (USC) power plant are:**

- Less fuel consumption,
- Highest power plant efficiency,
- Less $CO_2$ emission,
- Better availability,
- Less $NO_x$, $SO_x$ and particular emissions,
- Large unit size possibility.

1. **Sub-critical:**

   **Pressure:** 221.1 bar

   **Temperature:** 538/538°C or 538/560°C

2. **Supercritical:**

   **Pressure:** 221.1 bar to 260 bar

   **Temperature:** 538/560°C to 566/593°C

3. **Ultra-Supercritical (USC)**

   **Pressure:** 260 bar

   **Temperature:** 593/593°C to 600/620°C

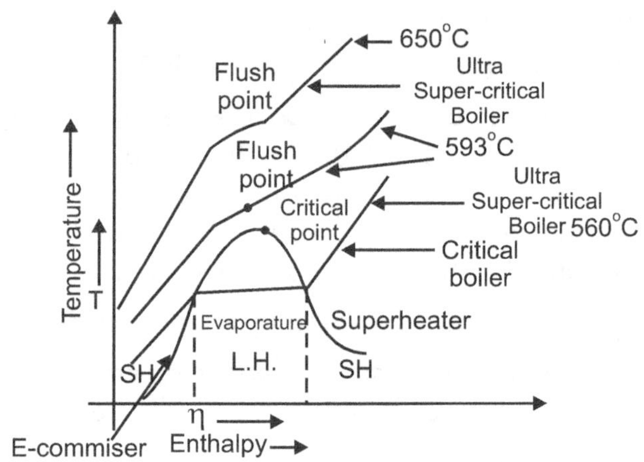


**Fig. 2.4**

**Table 2.1: Sub-critical Vs. Supercritical Vs. Ultra Supercritical Boilers**

| Description | Unit | Subcritical 16.7 Mpa 5.38/5.38 deg. C | Supercritical 24.1 Mpa 566/593 deg. C | Ultra Supercritical 24.1 Mpa 600/600 deg. C | Reduction Sub Vs SC | Reduction Sub Vs USC |
|---|---|---|---|---|---|---|
| Turbine heat rate | kCal./kWh | 1918 | 1838 | 1814 | 80(4.2%) | 104(5.5%) |
| Boiler efficiency | % | 87% | 87% | 87% | 0 | 0 |
| Plant heat rate | kCal./kWh | 2207 | 2113 | 2085 | 94(4.2%) | 121(5.5%) |
| Coal consumption | MMT/Annum | 3.676 | 3.524 | 3.477 | 0.152(4.2%) | 0.199(%.5%) |
| Ash generation | MMT/Annum | 1.544 | 1.48 | 1.46 | 0.064(4.2%) | 0.084(5.5%) |
| $CO_2$ | MMT/Annum | 7.278 | 6.981 | 6.888 | 0.297(4.2%) | 0.39(5.5%) |
| $SO_x$ | MMT/Annum | 0.0230 | 0.0221 | 0.0218 | 0.0009(4.2%) | 0.0012(5.5%) |

## 2.3 HYPERBOLIC COOLING TOWER

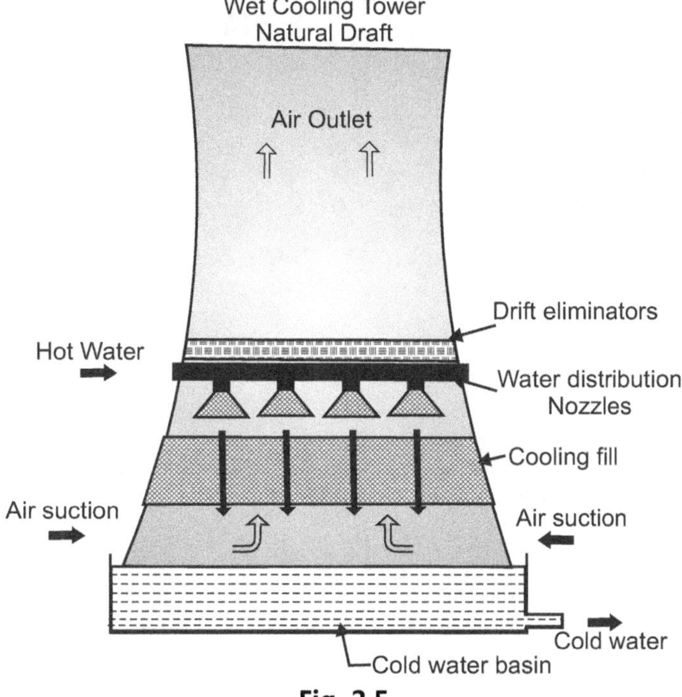


**Fig. 2.5**

Hyperboloid (sometimes known as hyperbolic) cooling towers have become the design standard for all natural-draft cooling towers.

- It is defined as density difference between the atmospheric air and the hot gas in the chimney.
- Purpose of a **cooling tower** is to reduce the temperature of circulating hot water to reuse this water again into the boiler, this hot water is coming from the condenser.
- The shape is like hyperbolic so it is called hyperbolic cooling tower.
- The hyperbolic shape also aids in accelerating the upward convective air flow, improving cooling efficiency.

**Working:**

Hot water is coming at the inlet of the tower and pumped-up to the header. The header contains nozzles and sprinklers which is used to spray water, and it will increase the surface area of water, after that water comes to cooling fill where the speed of water is reduced. Because of slow speed and more contact area of water, it makes a good connection between air and hot water. The air moves from bottom to the top due to density difference. The process will reduce the temperature of water by evaporation process and cooled water is collected at the bottom of the cooling tower, and this cooled water is used again in the boiler.

APPROACH = Cold water temperature – Wet bulb temperature

RANGE = Hot water temperature – Cold water temperature

$$\text{COOLING TOWER EFFICIENCY} = \frac{\text{Range}}{(\text{Range} + \text{Approach}) \times 100}$$

**Effectiveness:** This is the ratio between the range and the ideal range (in percentage), i.e. difference between cooling water inlet temperature and ambient wet bulb temperature.

$$\text{Effectiveness} = \frac{\text{ange}}{(\text{Range} + \text{Approach})} \cdot$$

The higher this ratio, the higher the cooling tower effectiveness.

**List of Commerciality Viable Waste Heat Recovery Devices:**

1. Waste heat recovery boilers.
2. Recuperators.
   (a) Metallic radiation recuperators.
   (b) Radiative recuperator.
   (c) Convective Radiative Recuperator.
   (d) Ceramic recuperators.
3. Regenerators.
4. Hot Wheel Exchanger.
5. Thermal Wheel.
6. Economizer.
   (a) Shell and tube heat exchanger.
   (b) Plate heat exchanger.
7. Heat Pump.
8. Heat Pipe.
9. Thermo-Compressor.
10. Direct contact heat Exchanger.
    (a) Hot-Hot Mixing.
    (b) Different Phase Mixing.
    (c) Hot Cold Mixing.

## 2.4 WASTE HEAT RECOVERY IN PROCESS INDUSTRIES

In Power generation and process industries continuous produce large amounts of hot waste gases and then "dumped" into the environment even though it could still be reused for some useful and economic purpose. These waste gases have been practiced for a long time form Generation of steam or power. The waste gases temperature can be at high, medium, or low at ~1100, 500, or 300°C, respectively. The gases meaningful calorific values (CV) are burnt with or without an additional support fuel. Several waste gases produced at high temperatures or in large volumes that contain considerable heat but no combustible fuel only need to be cooled (and at times cleaned) before release into the atmosphere.

**Waste heat boilers:**

The devices that cool the gases and produce steam are generally called waste heat boilers. The supplementary fuel may or may not be required.

The two types are:

1. Heat recovery steam generators (HRSGs) in power plants.
   - Gas turbines (GTs) exhaust large amounts of clean, medium hot gases used in boilers, usually called HRSGs or turbine exhaust gas (TEG) boilers.
2. Waste heat recovery boilers (WHRBs) in process industries.
   - In metallurgical, cement, and similar plants, large amounts of dusty hot gases generated are cleaned and cooled, steam is produced in boilers generally known as Waste heat recovery boilers (WHRBs) in process industries.
   - Waste heat recovery boilers (WHRBs) in process industries are bulky and expensive.
   - The cost of steam and power generated is higher.

To satisfy the condition of a waste heat system the following conditions are met:

1. The cost of the regular fuels is high and hence the high-cost power and steam from waste heat boilers are acceptable.

2. The process demands its inclusion.

## Waste Heat Recovery Boilers (WHRB):

Waste heat boilers are ordinary water tube boilers in which the hot gases from gas turbines pass over a number of parallel tubes containing water. The water is vaporized in the tubes and collected in the steam drum from which it is drawn off for use as heating or processing steam.

Because the exhaust gas are usually in the medium temperature range and in order to to conserve space, a more compact boiler can be produced if the water tubes are finned in order to increase the effective heat transfer area on the gas side the figure shows a mud drum set off tubes over which the hot gases make a double pass, and a steam drum which collects team generated above the water surface the pressure at which the steam is generated and the rate of steam production depends on the the temperature of of waste heat the pressure of a pure vapour in the presence of it's liquid is a function of the temperature of the liquid from which it is evaporated.

If the waste heat it in the exhaust gases is insufficient for generating the required amount of process steam, auxiliary Burners which burn fuel in the waste heat boiler or an after-burner in the exhaust flue gases are added. waste heat boilers build in capacities from 25 m$^3$ almost 30000 m$^3$/min of exhaust gas. Waste heat boilers are available in a variety of capacitors allowing for gas Intex from 1000 to 1 million ft 3/min. Water tube boilers that use medium to high temperature exhaust gases to generate steam in case where the waste heat is not sufficient for producing desire levels of steam auxiliary burners for an afterburner can be added to attain higher steam output the steam can be used for process heating or for power generation.

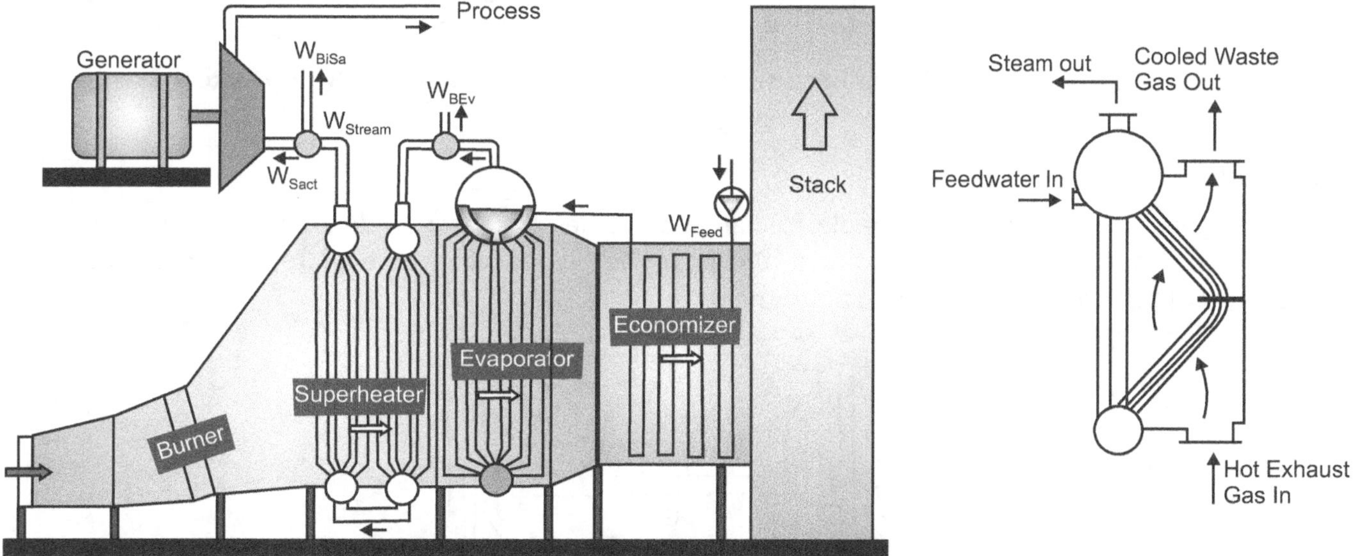


**Fig. 2.6: Waste Heat Recovery Boilers (WHRB)**

## Recuperators:

The recuperator is a heat exchange device that takes place between the flue gases and the air through metallic or ceramic walls. Duct or tubes carry the air for combustion to be pre-heated, the other side contains the waste heat stream. A recuperator is used for recovering waste heat from flue gases.

## Types of Recuperator:

(a) Metallic radiation recuperators.

(b) Radiative recuperator.

(c)  Convective Radiative Recuperator.

(d)  Ceramic recuperators.

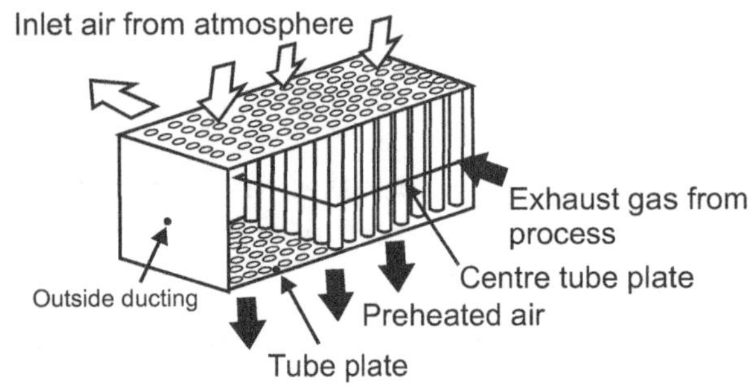


**(a) Waste Heat Recovery using Recuperator**

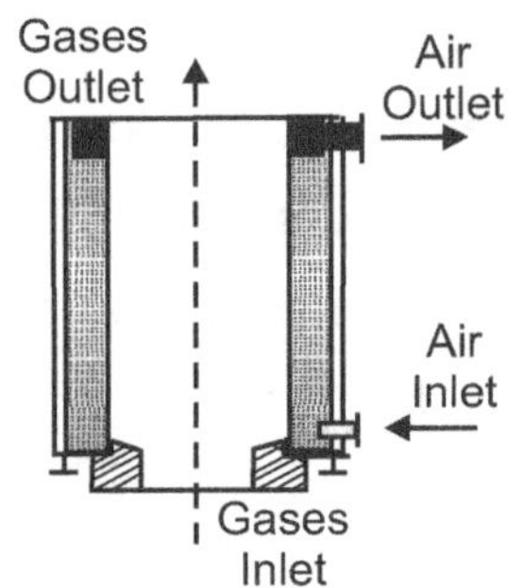


**(b) Metallic Radiation Recuperator**

**Fig. 2.7**

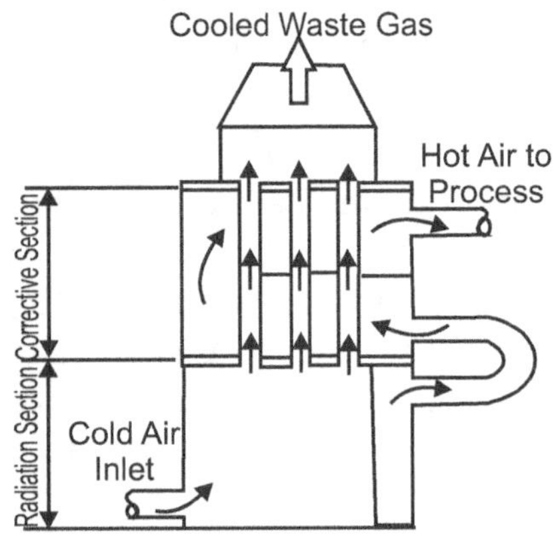


**(c) Convective Radiative Recuperator**

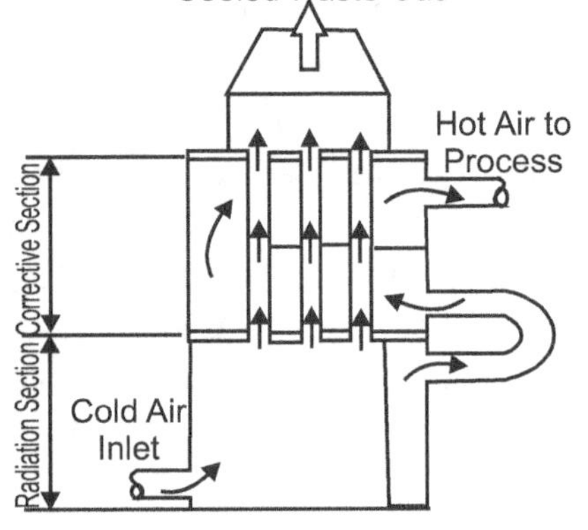
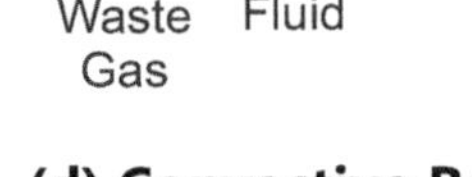


**(d) Convective Recuperator**

**Fig. 2.8**

## Regenerator

The Regeneration which is preferable for large capacities has been widely used in glass and steel melting furnaces. Important relations exist between the size of the regenerator, thickness of brick, conductivity of brick and heat storage ratio of the brick.

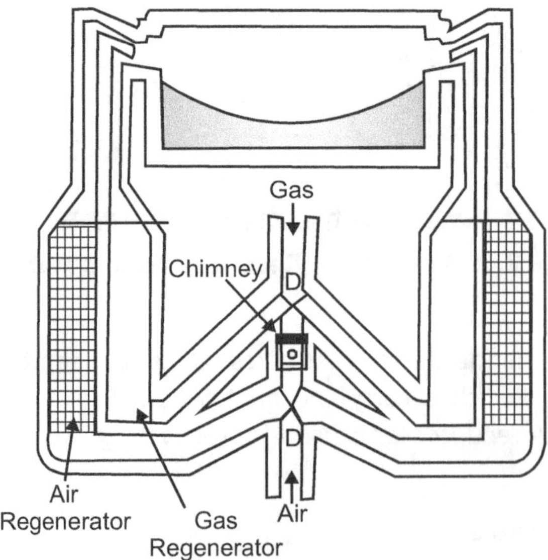


**Fig. 2.9: Regenerator**

## Hot Wheel Exchanger:

A thermal wheel, also known as a rotary heat exchanger, or rotary air-to-air enthalpy wheel, or heat recovery wheel, is a type of energy recovery heat exchanger positioned within the supply and exhaust air streams of an air-handling system or in the exhaust gases of an industrial process. A heat wheel is a growing applications in low to medium temperature waste heat recovery systems.

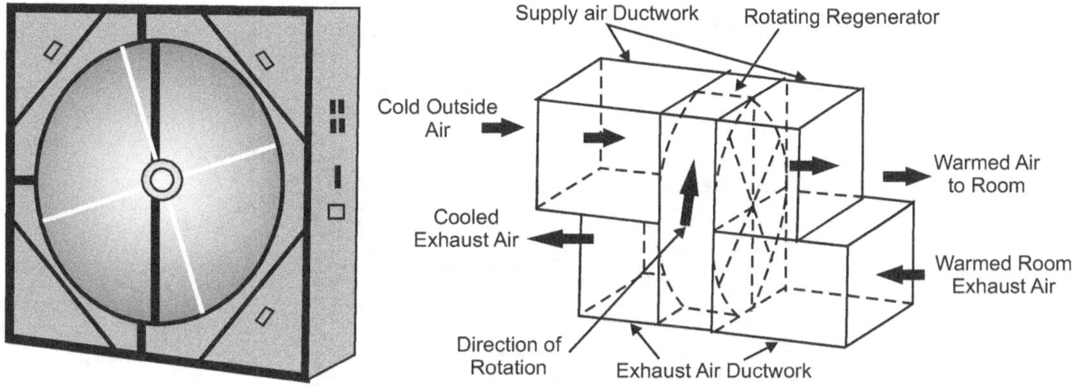


**Fig. 2.10: Hot Wheel Exchanger**

## Heat Pipe

The heat pipe is a thermal energy absorbing and transferring system. In heat pipe there is no moving parts and hence it requires minimum maintenance. A heat pipe can transfer up to 100 times more thermal energy than copper, the best known conductor.

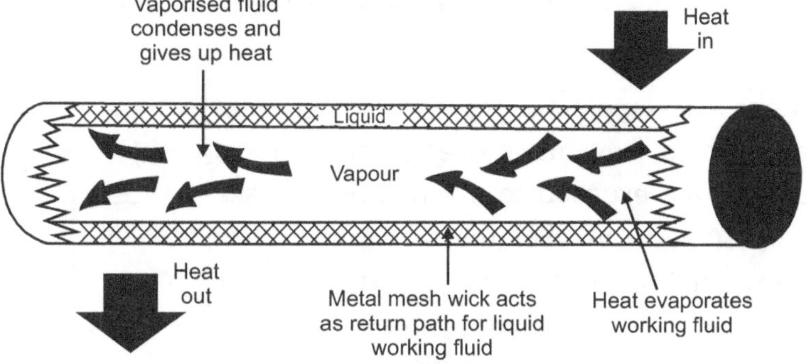


**Fig. 2.11: Heat Pipe**

## Economizer

Economizers are mechanical devices intended to reduce energy consumption, or to perform useful function such as preheating a fluid. The term economizer is used for other purposes as well. Boiler, power plant, heating, refrigeration, ventilating, and air conditioning In case of boiler system, economizer can be provided to utilize the flue gas for preheating the boiler feed water.

For every 22°C reduction in flue gas temperature through an economiser or a preheater, there is 1% saving of fuel in the boiler.

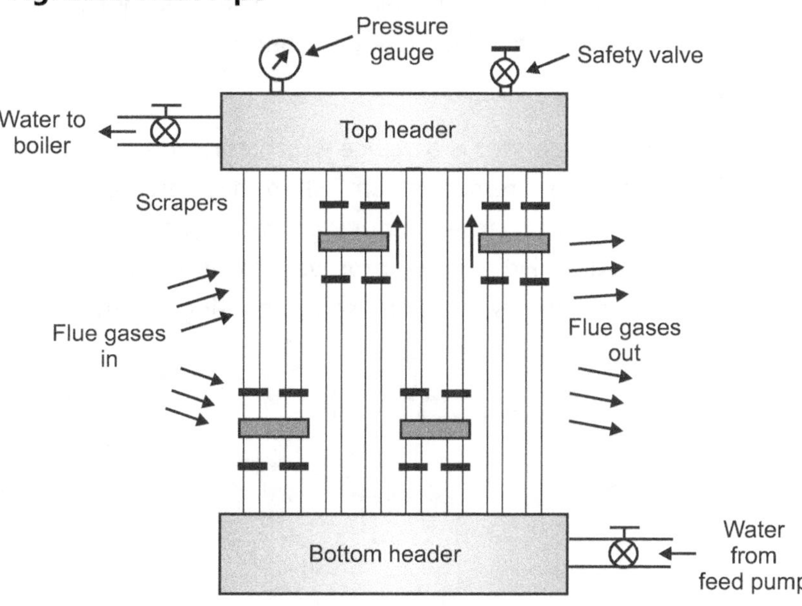


**Fig. 2.12: Economiser**

### Shell and Tube Heat Exchanger:

A shell and tube heat exchanger is the most common type of heat exchanger in oil refineries and other large chemical processes, and is suited for higher-pressure applications. As its name implies, this type of heat exchanger consists of a shell with a bundle of tubes inside it. The medium containing waste heat is a liquid or a vapour which heats another liquid, then the shell and tube heat exchanger must be used since both paths should be sealed to contain the pressures of their respective fluids.

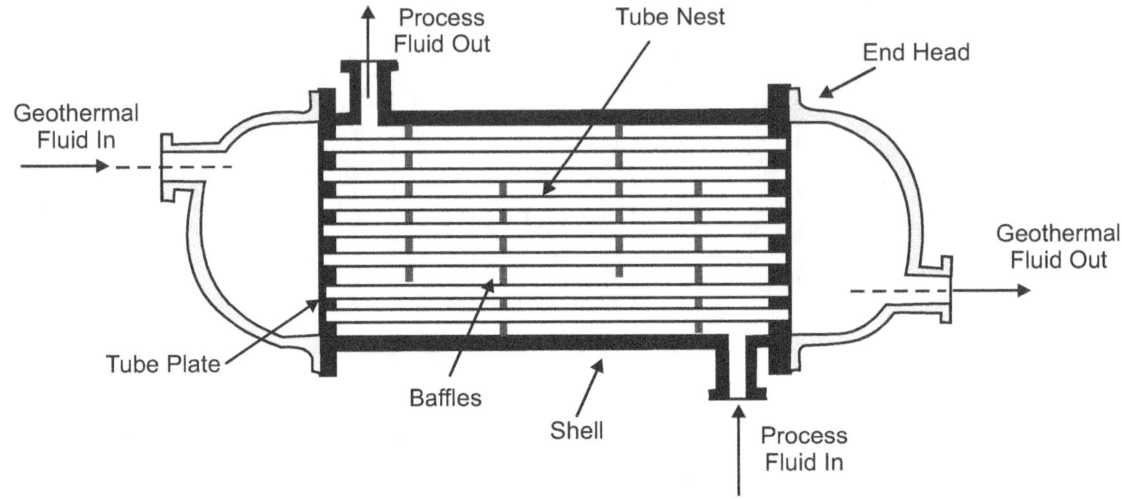


**Fig. 2.13: Shell and Tube Heat Exchanger**

### Plate Heat Exchanger

A plate heat exchanger uses metal plates to transfer heat between two fluids. This is a major advantage over a conventional heat exchanger that, the fluids are exposed to much larger surface area because the fluids are spread out over the plates. the plate type heat exchanger, consists of a series of separate parallel plates forming thin flow passes. Each plate is separated by gaskets and the hot stream passes parallelly through alternative plates, the liquid which is to be heated flows in parallel between the hot plates. To improve heat transfer the plates are corrugated.

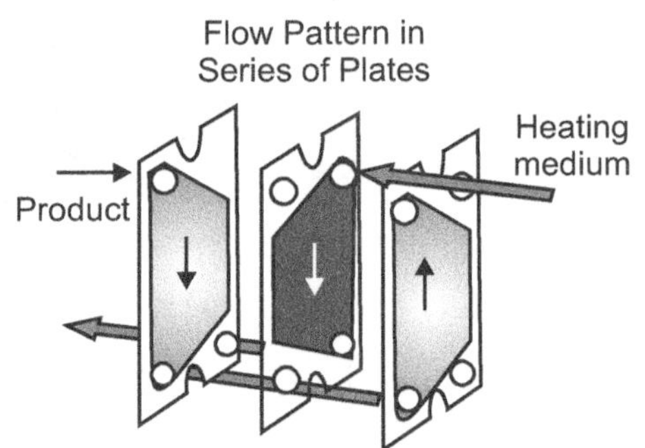


**Fig. 2.14: Plate heat exchanger**

### Thermocompressor :

The thermocompressor is a device which converts low pressure steam to a very high pressure steam and reuses it as a medium pressure steam, where high pressure HP steam is accelerated into a high velocity fluid through a nozzle. The large amount of steam energy is in the form of latent heat and thus thermo-compressing would give a large improvement in waste heat recovery. It is used in evaporators where the boiling steam is recompressed and used as heating steam.

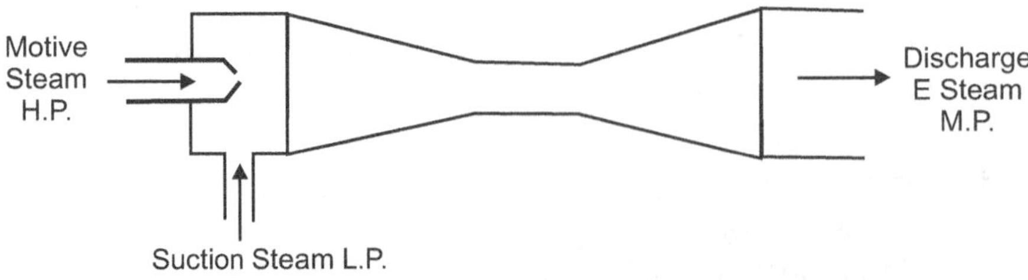


**Fig. 2.15: Thermocompressor**

**Direct Contact Heat Exchanger :**

In Direct Contact Heat Exchanger Low pressure steamis used to preheat the feed water or some other fluid. Direct Contact Heat Exchanger use in a steam generating station.

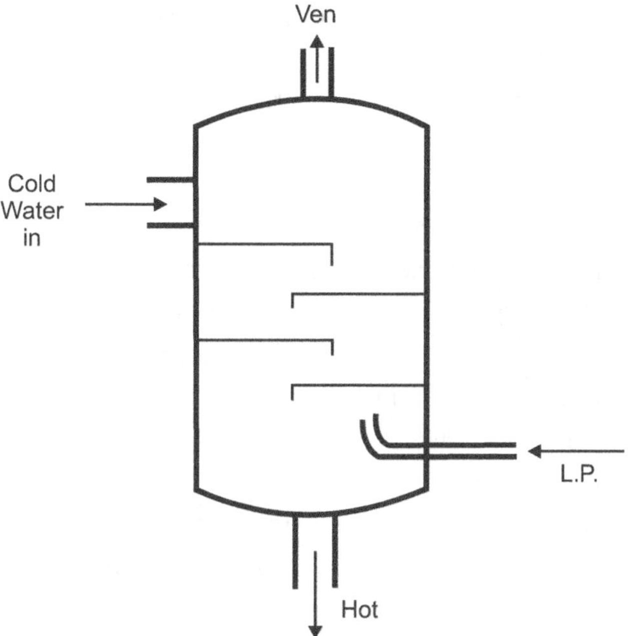


**Fig. 2.16: Direct Contact Condenser**

## Multiple Choice Questions

1. Latent heat of steam at lower pressure is lower ......
   (a) True              (b) False

2. The heat which is required to change the phase from water at 100°C to saturated steam is called ......
   (a) Latent Heat      (b) Sensible Heat      (c) Super Heat      (d) Specific Heat

3. Ideal trap for steam mains is ......
   (a) Thermodynamic      (b) float      (c) inverted bucket      (d) bimetallic

4. The best quality of steam for industrial process heating is ......
   (a) Dry saturated      (b) Super-heated      (c) Wet Steam      (d) High pressure steam

5. Cooling tower effectiveness is the ratio of ......
   (a) Range/(Range + Approach)      (b) Range/Approach
   (c) Approach/(Range + Approach)      (d) Approach/Range

6. Cooling tower reduces the water temperature close to ......
   (a) Dry bulb temperature      (b) Dew point temperature
   (c) Ambient wet bulb temperature      (d) None of the above

7. Which type of cooling towers with maximum heat transfer between air to water is ......
   (a) Natural draft      (b) Mechanical draft      (c) Both (a) and (b)      (d) None of the above

8. LG ratio in a cooling tower is the ratio of ......
   (a) Length and girth      (b) Length and temperature gradient
   (c) Water flow rate and their mass flow rate      (d) Air mass flow rate and water flow rate

9. Generally which Type of the steam use in process industries ......
   (a) Dry saturated      (b) Super-heated      (c) Wet Steam      (d) High pressure steam

10. Steam Temperature of ultra-supercritical boiler (USC) in °C
    (a) 520  (b) 620  (c) 560  (d) 330

11. What is the working pressure of ultra-supercritical boiler in bar?
    (a) More than 260 bar  (b) Less than 260 bar  (c) Both (a) and (b)  (d) None of the above

12. What does WHRB stands for?
    (a) Waste heat recovery boiler.  (b) World heat recovery boiler.
    (c) Waste harvest recovery boiler.  (d) All of the above.

13. What is the efficiency of ultra-supercritical boiler?
    (a) 38 to 42%  (b) 42 to 46%  (c) 46 to 48%  (d) None of the above

14. Which of the following are not waste heat recovery device?
    (a) Direct contact heat exchanger.  (b) Economizer.
    (c) Radiator.  (d) Thermo compressor.

15. Which of the following is not the type of recuperator?
    (a) Metallic radiation  (b) Radiative  (c) Ceramic  (d) Shell and tube type

16. Major advantage of waste heat recovery in industry is ......
    (a) Reduction in pollution.  (b) Increase in efficiency.
    (c) Both (a) and (b).  (d) None of the above.

17. Identify the Process B to C in the Fig. 2.15.

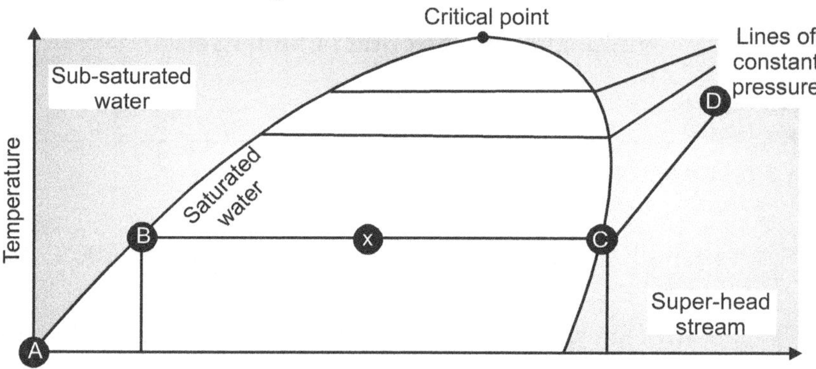


**Fig. 2.17**

    (a) Sensible heat  (b) Latent heat  (c) Both (a) and (b)  (d) None of the above

18. Recuperator is used mainly as waste heat recovery system in a ......
    (a) Boiler  (b) Compressor
    (c) Billet reheating furnace  (d) None of the above

19. Heat recovery equipment will be the most effective when the temperature of flue gas is ......
    (a) 250°C  (b) 200°C  (C) 400°C  (d) 280°C

20. Air preheater is not used as waste heat recovery system in a ......
    (a) Boiler  (b) Billet reheating furnace
    (c) Compressor  (d) Heat treatment furnace

## Answers

| Question No. | 1 | 2 | 3 | 4 | 5 | 6 | 7 | 8 | 9 | 10 |
|---|---|---|---|---|---|---|---|---|---|---|
| **Answer** | (b) | (a) | (a) | (a) | (a) | (c) | (b) | (c) | (a) | (b) |
| **Question No.** | 11 | 12 | 13 | 14 | 15 | 16 | 17 | 18 | 19 | 20 |
| **Answer** | (a) | (a) | (c) | (c) | (d) | (c) | (b) | (a) | (c) | (c) |

# Chapter 3

# RECENT TRENDS IN MANUFACTURING INDUSTRY

**Weightage of Marks = 20, Teaching Hours = 14**

## Syllabus

3.1 **Smart Manufacturing Technology:** Introduction, Elements and Applications.

3.2 **Automation:** Need, Basic elements of automated systems, Automation principles and strategies, Benefits.

3.3 **Types of Automation:** Fixed, Programmable, Flexible, Hard and soft automation.

3.4 **Industrial Robotics:** Robot anatomy, Robot control systems, End effectors, Sensors in robotics, Industrial robot applications.

3.5 4-D Printing Technology - Printing Techniques, 3D scanning technology - function, Applications.

## About this Chapter

At the end of this chapter, students will be able to:

- List various elements of smart manufacturing.
- Interpret the automation in mechanical industry.
- List different types of automation.
- Select robot for given application.
- Compare 4D printing technology with 3D printing technology.
- Describe the importance of 3D scanning with reverse engineering.

## 3.1 SMART MANUFACTURING

**Definition of Smart Manufacturing:**

Smart manufacturing is the use of real-time data and technology when, where and in the forms that are needed by people and machines.

According to the National Institute of Standards and Technology (NIST) Smart Manufacturing are systems that are "fully-integrated, collaborative manufacturing systems that respond in real time to meet changing demands and conditions in the factory, in the supply network, and in customer needs."

It is also defined as, "**Smart Manufacturing is the ability to solve existing and future problems via an open infrastructure that allows solutions to be implemented at the speed of business while creating advantaged value.**"

**Elements of Smart Manufacturing:**

- Internet of Things (IoT)
- Digital Transformation
- Artificial Intelligence
- Cloud Computing
- Data Science and Analytic

## 1. Internet of Things (IoT):

The Internet of Things (IoT) is a system of interrelated computing devices, mechanical and digital machines, objects, animals or people that are provided with unique identifiers (UIDs) and the ability to transfer data over a network without requiring human-to-human or human-to-computer interaction.

The definition of the Internet of Things has evolved due to the convergence of multiple technologies, real-time analytics, machine learning, commodity sensors, and embedded systems.

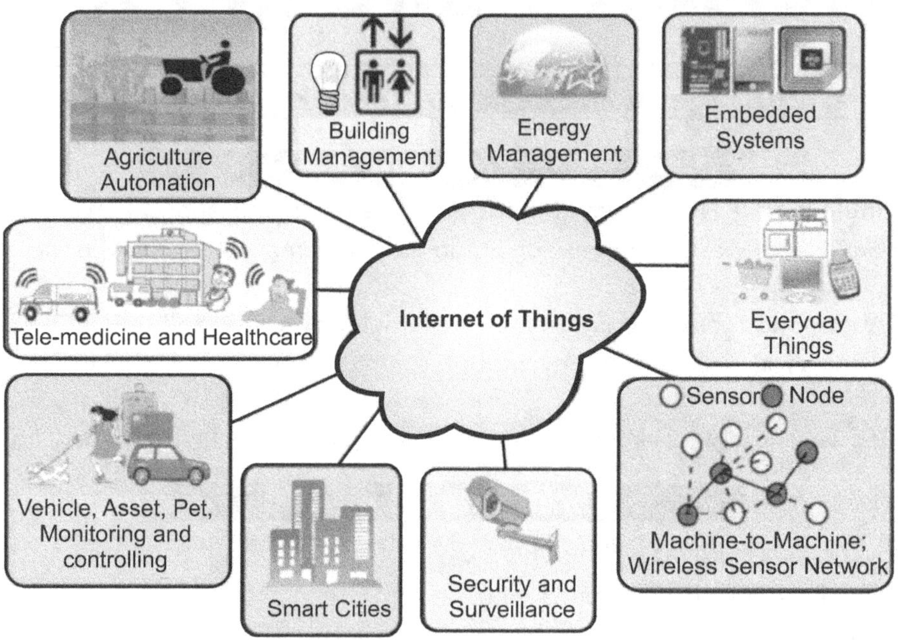


**Fig. 3.1: Internet of Things (IoT)**

## 2. Digital Transformation:

Digital transformation is the process of using digital technologies to create new or modify existing business processes, culture, and customer experiences to meet changing business and market requirements. This re-imagining of business in the digital age is digital transformation.

It transcends traditional roles like sales, marketing, and customer service. Instead, digital transformation begins and ends with how you think about, and engage with, customers. As we move from paper to spreadsheets to smart applications for managing our business, we have the chance to reimagine how we do business how we engage our customers with digital technology on our side.

For small businesses just getting started, there's no need to set up your business processes and transform them later. You can future-proof your organization from the word go. Building a 21st-century business on stickies and handwritten ledgers just isn't sustainable. Thinking, planning, and building digitally sets you up to be agile, flexible, and ready to grow.

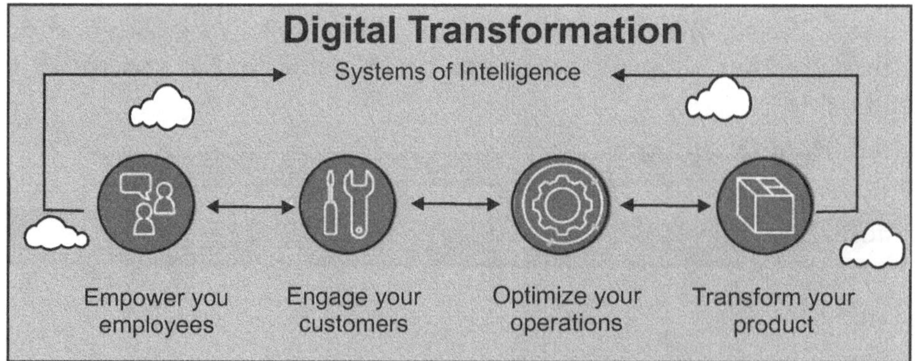


**Fig. 3.2: Digital Transformation**

## 3. Artificial Intelligence:

Artificial intelligence (AI) is an area of computer science that emphasizes the creation of intelligent machines that work and react like humans. Some of the activities computers with artificial intelligence are designed for include: **Speech recognition, Learning, Planning and Problem solving.**

Artificial intelligence is a branch of computer science that aims to create intelligent machines. It has become an essential part of the technology industry.

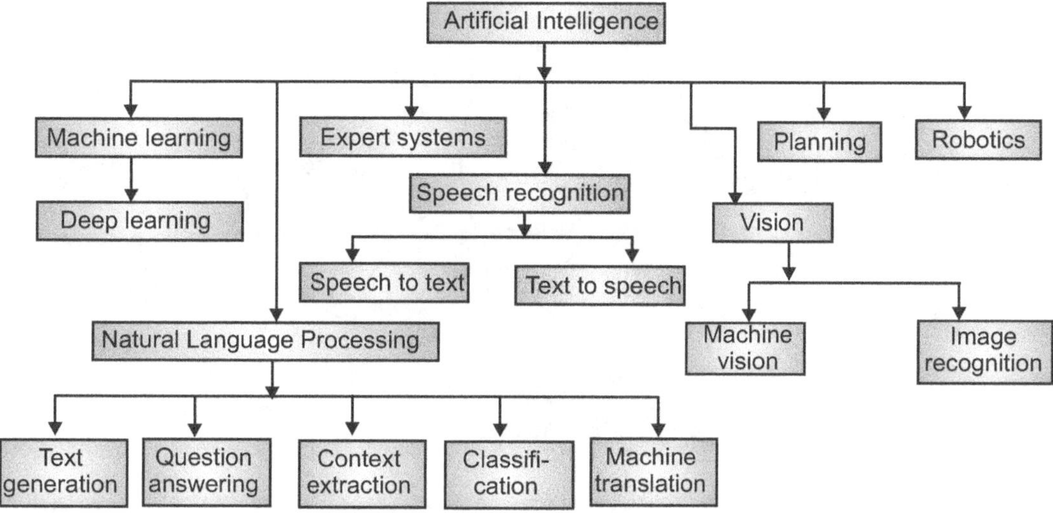


**Fig. 3.3: Artificial Intelligence**

## 4. Cloud Computing:

Cloud computing is the on-demand availability of computer system resources, especially data storage and computing power, without direct active management by the user. The term is generally used to describe data centers available to many users over the Internet. Large clouds, predominant today, often have functions distributed over multiple locations from central servers. If the connection to the user is relatively close, it may be designated an edge server.

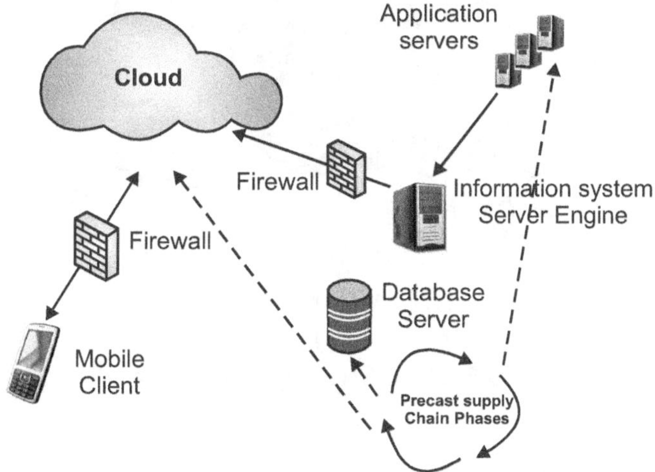


**Fig. 3.4: Cloud Computing**

## 5. Data Science and Analytic :

Data science is a concept used to tackle big data and includes data cleansing, preparation, and analysis. A data scientist gathers data from multiple sources and applies machine learning, predictive analytics, and sentiment analysis to extract critical information from the collected data sets. They understand data from a business point of view and can provide accurate predictions and insights that can be used to power critical business decisions.

Data science, analytics, and machine learning are growing at an astronomical rate and companies are now looking for professionals who can sift through the goldmine of data and help them drive swift business decisions efficiently. IBM predicts that by 2020, the number of jobs for all U.S. data professionals will increase by 364,000 openings to 2,720,000. We caught up with Eric Taylor, Senior Data Scientist at Circle Up, in a Simply learn Fireside Chat to find out what makes data science and data analytics such an exciting field and what skills will help professionals gain a strong foothold in this fast-growing domain.

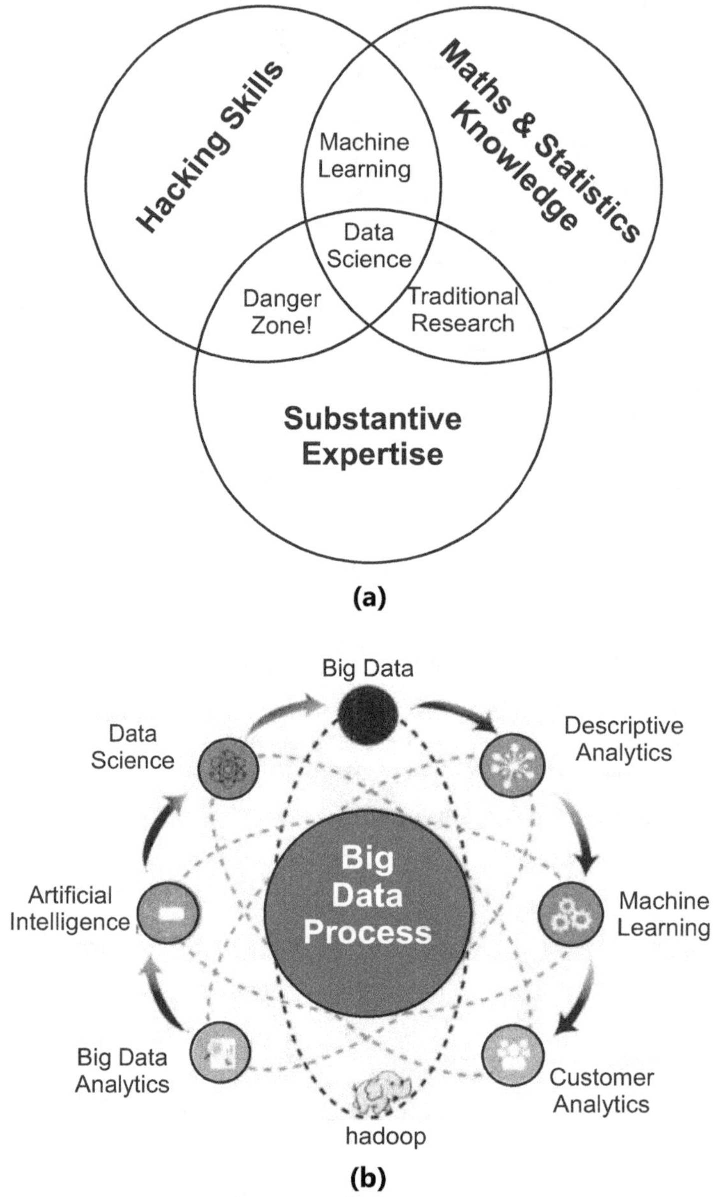


**Fig. 3.5: Data Science and Analytic**

**Applications of Smart Manufacturing**

**Kuka – IoT enabled factory**

- A German robotics maker.
- Build a Lot enabled factory.
- The factory has hundred robots.
- Robots are connected with a private cloud.
- 800 cars are manufactured per day.

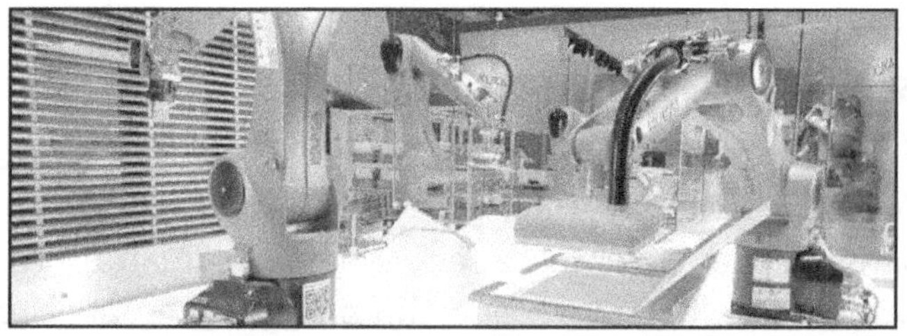

**Fig. 3.6**

### DeWalt Construction internet of things

- A tool manufacturer.
- Launched construction internet of things initiative.
- Uses IoT platform and Wi-Fi mesh network.
- Tracks workers and equipments.
- Monitors sites as large as a football stadium.

Fig. 3.7

### ABB - YuMi

- A power and robotics firm.
- Monitors robots via connected sensors.
- Preventive maintenance.
- YuMi model.
- An initiative collaboration between humans and robots.

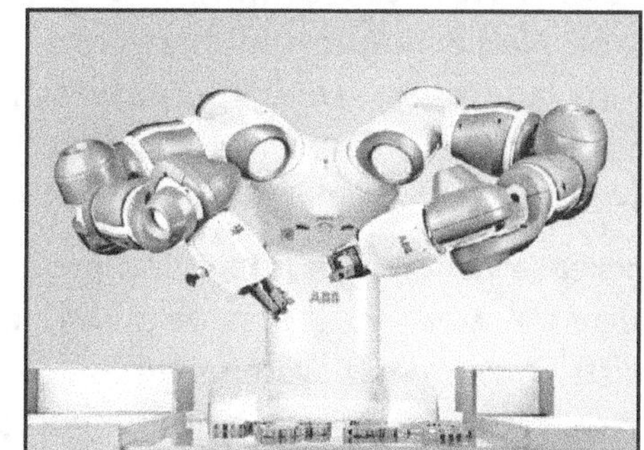

Fig. 3.8

### Amazon - Robotics Shelves

- An E- commerce company.
- Uses robotic shelves.
- Robots carry and rearrange shelves.
- Automated product search.
- Robots locate and bring shelves to workers.

Fig. 3.9

### Caterpillar - AR App:

- A heavy equipment maker.
- Uses argument reality (AR) with IoT.

Fig. 3.10

## 3.2 INTRODUCTION OF AUTOMATION

The word 'Automation' is derived from greek words "Auto" (self) and "Matos" (moving). Automation therefore is the mechanism for systems that "move by itself". However, apart from this original sense of the word, automated systems also achieve significantly superior performance than what is possible with manual systems, in terms of power, precision and speed of operation.

Automation is a set of technologies that results in operation of machines and systems without significant human intervention and achieves performance superior to manual operation

Automation is the technology by which a process or procedure is accomplished without human assistance.

**"Automation = Automatic Control"**

## Need of Automation:

1. **Reduce Worker Fatigue and Effort or Labor Intensive Operation:** Typically, humans dislike banal, repetitive tasks. However, computer systems perform them without complaint. Tasks that lack variability provide a place for automated systems to shine, but this also holds true for systems utilizing advanced sensors and integration. If the task requires conditions not suited to human comfort or focus, consider automation.

2. **Prevent Products or Materials from Being Damaged or Destroyed:** Humans make mistakes when they fatigue. This embodies the sentiment of the "human condition." Mistakes using tools mean damaging raw materials, components, assemblies, and end products.

3. **Prevent Non-conforming Product from Shipping:** Computers controlling robots do not forget steps. Neglecting to put in a screw requires a human touch. A machine not doing it yields an error to be addressed. Does the process require doing something in a specific order to improve yield? Automated systems will not violate the instruction set. Moreover, automated systems may employ inspection capabilities. Tune the system and allow the data to roll in without preference or bias.

4. **Increase Efficiency:** Improving processes for efficiency makes a company more competitive, but do people always do the same thing, in the same way, every time they do it ? No, human variation exists. Automated systems allow for improvements that benefit from consistent execution. Perfect planning and training do not defend against the human touch.

5. **Collect Better Data:** Remove the accidental data entry or missed data point from logging. Make the method of collecting sensor and process data regulated.

6. **Collect Better Data:** Remove the accidental data entry or missed data point from logging. Make the method of collecting sensor and process data regulated.

7. **Devise the Right Process Improvements:** Automated systems now collect reliable data. The database provides a searchable forum. What comes next? Equipped with copious amounts of reliable data, engineers make the most of this information. Where problems existed, light shines on the problem. Rather than just changing to seek "continuous improvement," make changes with better information.

8. **Devise the Right Process Improvements:** Automated systems now collect reliable data. The database provides a searchable forum. What comes next? Equipped with copious amounts of reliable data, engineers make the most of this information. Where problems existed, light shines on the problem. Rather than just changing to seek "continuous improvement," make changes with better information.

## Basic elements of an automated system:

1. **Power:** To accomplish the process and operate the automated system.

2. **Program of instructions:** To direct the process.

3. **Control system:** To actuate the instructions.

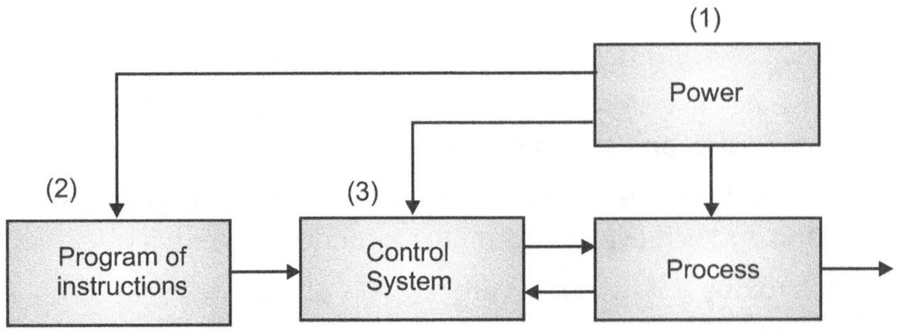


**Fig. 3.11**

## Power Source:

### Power for the process:

- To drive the process itself.
- To load and unload the work unit.
- Transport between operations.

### Power for automation:

- Controller unit.
- Power to actuate the control signals.
- Data acquisition and information processing.

## Program of Instructions:

Set of commands that specify the sequence of steps in the work cycle and the details of each step:

- **Example:** CNC part program.
- During each step, there are one or more activities involving changes in one or more process parameters.
- **Examples:**
  - Temperature setting of a furnace.
  - Axis position in a positioning system.
  - Motor on or off.

## Control System:

## Closed-loop (feedback) control system –

A system in which the output variable is compared with an input parameter, and any difference between the two is used to drive the output into agreement with the input.

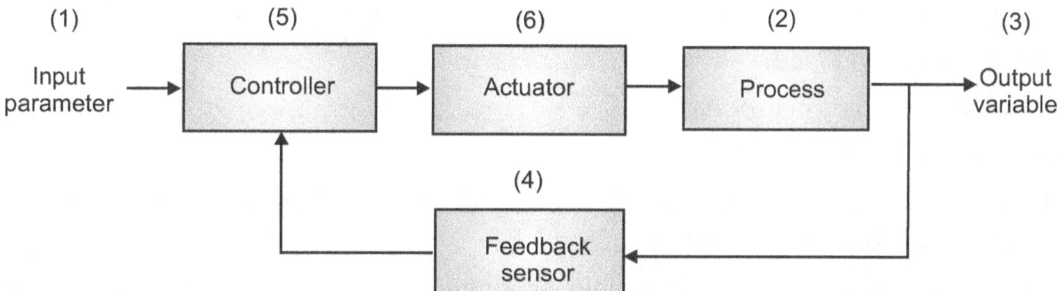


**Fig. 3.12: Closed Loop Control System**

## Open-loop control system

Operates without the feedback loop, Simpler and less expensive. Risk that the actuator will not have the intended effect.

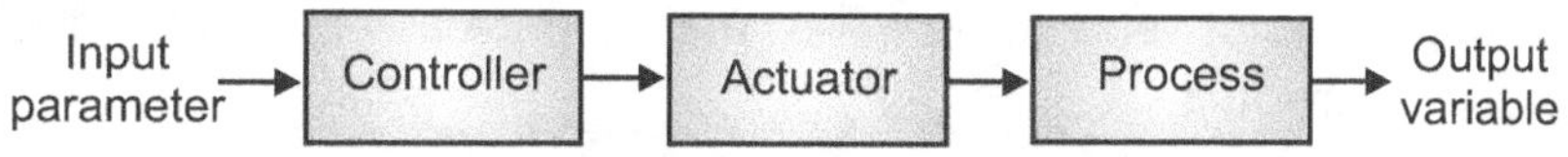


**Fig. 3.13: Open Loop Control System**

## Principle of Automation:

The Principle is a common sense approach to automation and process improvement projects. U means "understand the existing process," S stands for "simplify the process," and A stands for "automated the process". It is common Sense approach to automation and process improvement. Similar procedures have been suggested in the manufacturing and automation trade literature, but none has a more captivating title than this one.

**U :** understand the existing process.

**S :** simplify the process.

**A :** automate the process.

## U: Understanding the Process :

- Study current process in detail.
- Find answers:
  1. What are inputs?
  2. What are outputs?
  3. What exactly happens between input and output ?
  4. What is function of process ?
  5. How does it add value to the product ?
  6. What are sequence of operations ?

## Simplify and Automate :

- Simplify the Process:

  (a) Simplify the existing process.

  (b) Generate answers to queries:
  1. What is the purpose of each step and transport?
  2. Is this step necessary?
  3. Can this step be eliminated?
  4. Does this step uses the most appropriate technology?
  5. How can the step be simplified?
  6. Can steps be combined?
  7. Automate the steps in process.

## Role of automation in industry:

Manufacturing processes, basically, produce finished product from raw/unfinished material using energy, manpower and equipment and infrastructure.

Since an industry is essentially a "systematic economic activity", the fundamental objective of any industry is to make profit.

$$\text{Roughly speaking, Profit} = \left(\frac{\text{Price}}{\text{Unit}} - \frac{\text{Cost}}{\text{Unit}}\right) \times \text{Production Volume}$$

So profit can be maximized by producing good quality products, which may sell at higher price, in larger volumes with less production cost and time. The major parameters that affect the cost/unit of a mass-manufactured industrial product.

## Strategies for Automation :

- **Specialization of operations:** Special purpose equipment to perform one operation with greatest possible efficiency.
- **Combined operations:** Reducing number of distinct production machines.
- **Simultaneous Operations:** Reducing total processing time.
- **Integration of operations:** Linking several workstation into a single integrated mechanism.
- **Increased flexibility:** To achieve max utilization of equipment.

- Improved material handling and storage: Reducing non-productive time.
- **On-line inspection:** Corrections to the process during manufacturing.
- Process control and optimization.
- Plant operations and control.
- CIM.

## Benefits of Automation:

**Lower Operating Costs:** Robots can perform the work of three to five people, depending on the task. In addition to savings on the cost of labour, energy savings can also be significant due to lower heating requirements in automated operations. Robots streamline processes and increase part accuracy, which means minimal material waste for your operation.

**Improved Worker Safety:** Automated cells remove workers from dangerous tasks. Your employees will thank you for safeguarding them against the hazards of a factory environment.

**Reduced Factory Lead Times:** Automation can keep your process in-house, improve process control and significantly reduce lead times compared to outsourcing or going overseas.

**Ability to be More Competitive:** Automated cells allow you to decrease cycle times and cost-per-piece while improving quality. This allows you to better compete on a global scale. Additionally, the flexibility of robots enables you to retool a cell to exceed the capabilities of your competition.

**Increased Production output:** A robot has the ability to work at a constant speed, unattended, 24/7. That means you've got the potential to produce more. New products can be more quickly introduced into the production process and new product programming can be done offline with no disruption to existing processes.

**Consistent and Improved Part Production and Quality:** Automated cells typically perform the manufacturing process with less variability than human workers. This results in greater control and consistency of product quality.

**Smaller Environmental Footprint:** By streamlining equipment and processes, reducing scrap and using less space, automation uses less energy. Reducing your environmental footprint can save real money.

**Better Planning:** Consistent production by robots allows a shop to reliably predict timing and costs. That predictability permits a tighter margin on most any project.

**Reduce Need for Outsourcing:** Automated cells have large amounts of potential capacity concentrated in one compact system. This allows shops to produce parts in-house that have previously been outsourced.

**Optimal Utilization of Floor Space:** Robots are designed on compact bases to fit in confined spaces. In addition to being mounted on the floor, robots can be mounted on walls, ceilings, rail tracks and shelves. They can perform tasks in confined spaces, saving you valuable floor space.

**Easy Integration:** Productivity will work with you to provide a complete system hardware, software and controls included. Your cell will be proven out at Productivity and shipped production-ready allowing you to start making parts as soon as it's installed in your shop.

**Maximize Labour:** Over the next three decades, statistics show that more than 76 million baby boomers will retire and only 46 million new workers will be available to replace them. During this time, your demand for labor will continue, making automation a real and viable solution.

### Increase Productivity and Efficiency:

- 24/7 production, JIT manufacturing-friendly.
- More uptime with historic efficiency figures above 90 percent.
- Secondary operations capability gauging, washing, deburring, etc.

- Real-time factory communications with automated cell and machines.
- Quick changeover for multiple parts, tooling and programs.
- Flexible multi-operations capability.

**Increase System Versatility:**

- System flexibility, easily retooled and repositioned for new production programs.
- Robots are flexible and can easily be redeployed in new applications.
- Robots have the ability to easily switch between a wide range of products without having to completely rebuild production lines.
- Quick changeover with auto grippers and vision allows for different part sizes and shapes to be part of the same run.
- Mixed-flow production approach allows for flexibility in adjusting to demand fluctuations.
- Robots are able to instantaneously "learn" new processes.
- Reduced changeover time.

**Home Automation System:**

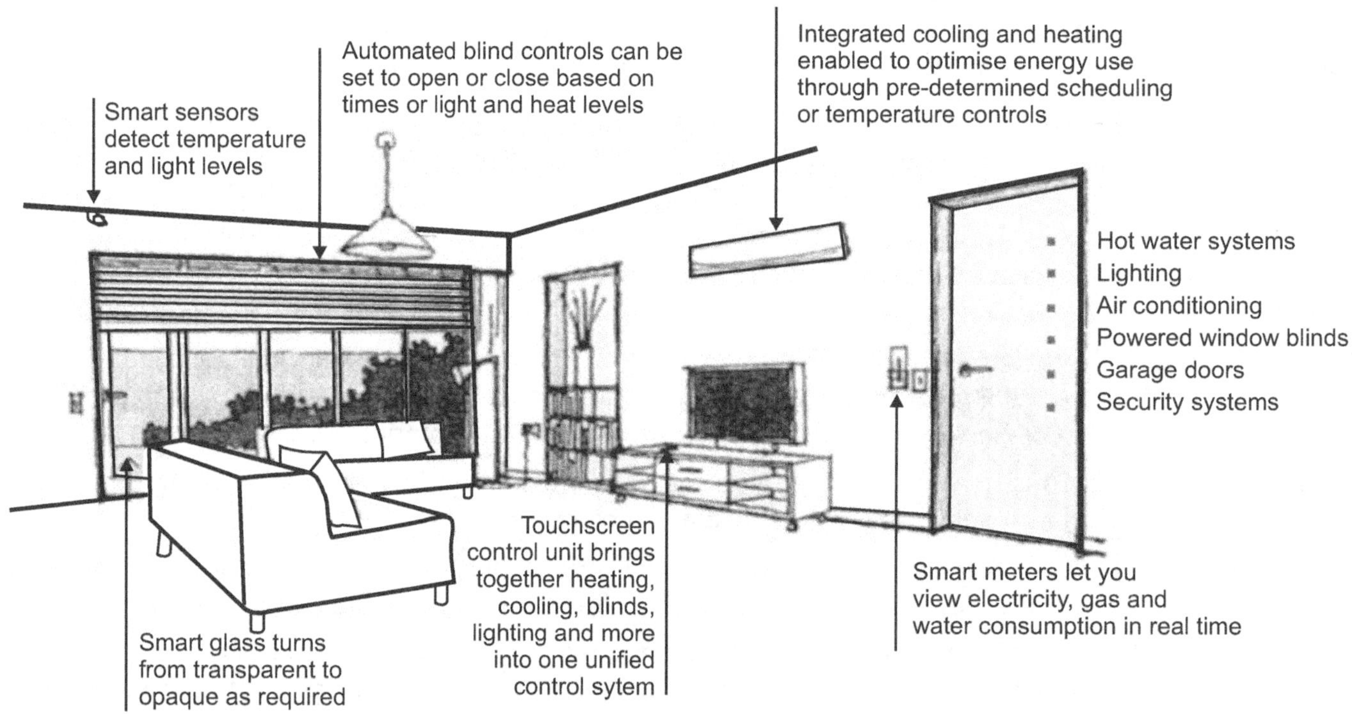


**Fig. 3.14**

## 3.3 TYPES OF AUTOMATION SYSTEMS

Automation systems can be categorized based on the flexibility and level of integration in manufacturing process operations. Various automation systems can be classified as follows:

**Fixed Automation:** It is used in high volume production with dedicated equipment, which has a fixed set of operation and designed to be efficient for this set. Continuous flow and Discrete Mass Production systems use this automation.

For example, Distillation Process, Conveyors, Paint Shops, Transfer lines etc. A process using mechanized machinery to perform fixed and repetitive operations in order to produce a high volume of similar parts.

- Sequence of processing operations is fixed by the equipment configuration.
- Sequence of simple operations.

- Integration and coordination of many operations in one equipment.
- Typical features:
  1. High initial cost.
  2. Custom engineered equipment.
  3. High production rates.
  4. Inflexibility.

**Programmable Automation:** It is used for a changeable sequence of operation and configuration of the machines using electronic controls. However, non-trivial programming effort may be needed to reprogram the machine or sequence of operations. Investment on programmable equipment is less, as production process is not changed frequently. It is typically used in Batch process where job variety is low and product volume is medium to high, and sometimes in mass production also.

For example, in Steel Rolling Mills, Paper Mills etc.

Production equipment is designed with capability to change sequence of operation to accommodate different product configurations.

- Operation sequence controlled by PROGRAM.
- Typical features:
  1. High investment in general purpose equipment.
  2. Low production rate.
  3. Flexibility to deal with product change.
  4. High suitability for batch production.
- Typical cycle for a product:
  1. Set up and reprogramming.
  2. Batch production of parts.
- **Examples:**
  1. NC Machine Tools.
  2. Industrial Robots.
  3. PLCs.

**Flexible Automation:** It is used in Flexible Manufacturing Systems (FMS) which is invariably computer controlled. Human operators give high-level commands in the form of codes entered into computer identifying product and its location in the sequence and the lower level changes are done automatically. Each production machine receives settings/instructions from computer. These automatically loads/unloads required tools and carries out their processing instructions. After processing, products are automatically transferred to next machine. It is typically used in job shops and batch processes where product varieties are high and job volumes are medium to low. Such systems typically use Multipurpose CNC machines, Automated Guided Vehicles (AGV) etc.

- Extension of programmable automation.
- Produce variety of parts with virtually no time lost for changeovers from one part style to next.
- No production time lost in set up and reconfiguration.
- System can produce various mixes of parts.
- Example: FMS.

**Typical Features:**
- High Investment for custom engineered equipments.

- Continuous production of mix of products.
- Medium production rates.
- Flexible to deal with product design variations.

## Hard Automation:

Hard automation is used for a specific production purpose where the processes are fixed. It is best suited for automated equipment that mass produced high volume products with few alterations or little changeovers. This type of automation has a high initial investment and high production rates, most typically automated assembly line machines.

The operations in a fixed sequence are usually simple and involve a plain linear or rotational motion, for example, or a combination of two such motions. Despite the simple operations involved, it is relatively difficult to make changes in the product design. Advantages of hard automation, according to SOMETECH, include low unit cost, automated material handling and a high production rate. The disadvantages are a high initial investment and the relative inflexibility in accommodating product changes.

## Soft Automation:

Soft automation can be used to produce a variety of parts with virtually no time lost for changeovers from one part style to another or for multiple small batches of a single product. No lost production time results when reprogramming the system. Advantages include continuous production of variable mixtures of product and flexibility to accommodate varying product designs. Disadvantages are a medium production rate, high long term production costs and a high unit cost compared with hard automation.

As Matt Charles wrote in a recent blog post on the website of industrial automation provider Cross Company, "What is Flexible Automation?," the Tesla Motors Factory in Freemont, CA utilizes soft automation. Although it builds this high-tech car, it could build just about any product. The "veritable robot army" features hundreds of robot arms that perform tasks such as metal-bending and assembly. Robot vehicles carry car chassis around and robot painting arms open and car doors to spray around them. The robots themselves typically have no sensing ability and rely solely on their mechanical repeatability. The overall system is programmed for flexible automation, though.

## Integrated Automation:

It denotes complete automation of a manufacturing plant, with all processes functioning under computer control and under coordination through digital information processing. It includes technologies such as computer-aided design and manufacturing, computer-aided process planning, computer numerical control machine tools, flexible machining systems, automated storage and retrieval systems, automated material handling systems such as robots and automated cranes and conveyors, computerized scheduling and production control. It may also integrate a business system through a common database. In other words, it symbolizes full integration of process and management operations using information and communication technologies. Typical examples of such technologies are seen in Advanced Process Automation Systems and Computer Integrated Manufacturing (CIM).

## Industrial Robotics:

Robot anatomy, robot control systems, end effectors, sensors in robotics, industrial Robot applications.

The term comes from a Czech word, robota, meaning "forced labour." The word robot first appeared in a 1920 play by Czech writer Karel Capek, R.U.R.: Rossum's Universal Robots. In the play, the robots eventually overthrow their human creators.

A robot is a re programmable multi-function manipulator designed to move material parts, tools or specialised devices, through variable programmed motions for the performance of a variety of tasks. (Robotic Institute of America, 1979).

Robots are devices that are programmed to move parts, or to do work with a tool. Robotics is a multidisciplinary engineering field dedicated to the development of autonomous devices, including manipulators and mobile vehicles. Roboticists develop man-made mechanical devices that can move by themselves, whose motion must be modelled, planned, sensed, actuated and controlled.

Robotics is, to a very large extent, all about system integration, achieving a task by an actuated mechanical device, via an "intelli shares with other domains, such as systems and control, computer science, character animation, machine design, computer vision, artificial intelligence, cognitive science, biomechanics, etc.

In addition, the boundaries of robotics cannot concepts and algorithms are being applied applications, and, vice versa, core technology from other domains (vision, biology, cognitive science or biomechanics, for example) are becoming crucial components in more and more modern robotic systems.

**Fig. 3.15: First Robot**

In 1954 George Devol invented the first digitally operated and a programmable robot called the Unimate. In 1956, Devol and his partner Joseph Engelberger formed the world's first robot company. In 1961, the first industrial robot, Unimate, went online in a General Motors automobile factory in New Jersey.

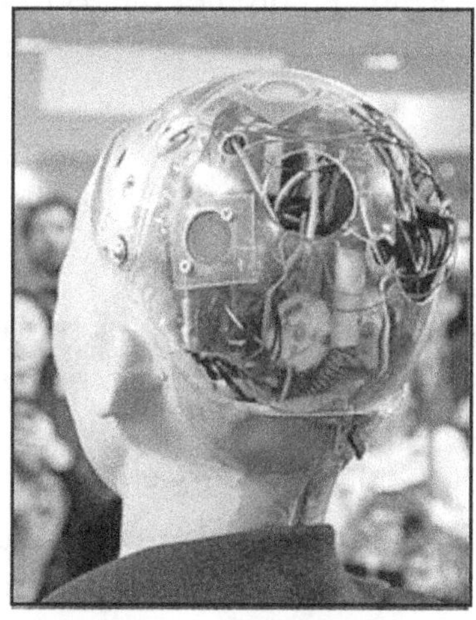

**Fig. 3.16: First Ever Robot Who got Citizenship. (Sophia)**

Sophia is a social humanoid robot developed by Hong Kong based company Hanson Robotics. Sophia was activated on February 14, 2016, and made her first public appearance at South by Southwest Festival (SXSW) in mid-March 2016 in Austin, Texas, United States. She is able to display more than 60 facial expressions.

Sophia has been covered by media around the globe and has participated in many high-profile interviews. In October 2017, Sophia became the first robot to receive citizenship of any country. In November 2017, Sophia was named the United Nations Development Programme's first ever Innovation Champion, and is the first non-human to be given any United Nation title.

Cameras within Sophia's eyes combined with computer algorithms allow her to see. She can follow faces, sustain eye contact, and recognize individuals. She is able to process speech and have conversations using a natural language subsystem. Around January 2018 Sophia was upgraded with functional legs and the ability to walk.

Sophia is conceptually similar to the computer program ELIZA, which was one of the first attempts at simulating a human conversation. The software has been programmed to give pre-written responses to specific questions or phrases, like a chatbot. These responses are used to create the illusion that the robot is able to understand conversation, including stock answers to questions like "Is the door open or shut?" The information is shared in a cloud network which allows input and responses to be analysed with blockchain technology.

David Hanson has said that Sophia would ultimately be a good fit to serve in healthcare, customer service, therapy and education. Sophia runs on artificially intelligent software that is constantly being trained in the lab, so her conversations are likely to get faster, Sophia's expressions are likely to have fewer errors, and she should answer increasingly complex questions with more accuracy.

**Types of Robots:**

**Industrial Robots:**

- It is defined as automatically controlled reprogrammable multipurpose manipulator designed to move in three or more axis, which may be either fixed in place or mobile for use in industrial automation applications.

- They do not get tired and they do not make errors associated with fatigue and so are ideally well suited to perform repetitive tasks.

- Some robots are programmed to faithfully carry out specific actions over and over again (repetitive actions) without variation and with a high degree of accuracy.

- Other robots are much more flexible as to the orientation of the object on which they are operating or even the task that has to be performed on the object itself, which the robot may even need to identify.

- A *collaborative robot or cobat* is a robot that can safely and effectively interact with human workers while performing simple industrial tasks.

**Mobile Robots:**

- Mobile robots have the capability to move around in their environment and are not fixed to one physical location. Mobile robots can be "autonomous" - autonomous mobile robot) which means they are capable of navigating an uncontrolled environment without the need for physical or electromechanical guidance devices.

- Alternatively mobile robots can rely on guidance devices that allow them to travel a pre-defined navigation route in relatively controlled space (.AGN' - autonomous guided vehicle). By contrast, industrial robots are usually more-or-less stationary.

**Humanoid Robot:**

- A humanoid robot is a robot with its body shape built to resemble that of the human body. A humanoid design might be for functional purposes, such as interacting with human tools and environments, for experimental purposes or for other purposes.

- In general, humanoid robots have a torso, a head, two arms, and two legs, though some forms of humanoid robots may model only part of the body,

- For example, from the waist up. Some humanoid robots may also have heads designed to replicate human facial features such as eyes and mouths.

- Researchers need to understand the human body structure and behavior (biomechanics) to build and study humanoid robots. On the other side, the attempt to the simulation of the human body leads to a better understanding of it.

- Besides the research humanoid robots are being developed to perform human tasks like personal assistance, where they should be able to assist the sick and elderly, and dirty or dangerous jobs. Regular jobs like being a receptionist or a worker of an automotive manufacturing line are also suitable for humanoids. In essence, since they can use tools and operate equipment and vehicles designed for the human form, humanoids could theoretically perform any task a human being can, so long as they have the proper software. However, the complexity of doing so is deceptively great.

- Humanoid robots, especially with artificial intelligence algorithms, could be useful for future dangerous and/or distant space exploration missions.

**Military Robots:**

- Military robots are autonomous robots or remote-controlled devices designed for military applications. It can be UAVs, Drones, mobile robots or even humanoid soldiers. Robotics are the future of warfare.

- Defence Research and Development Organisation (DRDO) plan to create super-intelligent robots to fight alongside human troops. Defense Research and Development Organization (DRDO) announced its intention to develop robotic soldiers for military applications.

- Daksh is a battery-operated remote-controlled robot on wheels and its primary role is to recover bombs. Developed by Defence Research and Development Organisation, it is fully automated. It has a shotgun, which can break open locked doors, and it can scan cars for explosives.

**Nancrobotics, (Nanobots):**

- Nanorobotics is the emerging technology field creating machines or robots whose components are at or close to the scale of a nanometre (10-9 meters). Nanomachines are largely in the research and development phase, but some primitive Nanobots.

- Researchers also hope to be able to create entire robots as small as viruses or bacteria, which could perform tasks on a tiny scale. Possible applications include micro surgery (on the level of individual cells), utility fog (a hypothetical collection of tiny robots that can replicate a physical structure), manufacturing, weaponry and cleaning.

**Bionics anti Biomimetics Robots:**

- Biomimetics or biomimicry is the imitation of the models, systems, and elements of nature for the purpose of solving complex human problems. Biomimcry seeks solutions to human challenges by emulating natures time-tested Patterns and strategies.

- Bionics is science of constructing artificial systems that have some of the characteristics of living systems. Bionics is distinct from bioengineering. which is the use of living things to perform certain industrial tasks

- These are used to apply the way animals move to the design of robots. BionicKangaroo was based on the movements and physiology of kangaroos.

- **Bionicngaroo:** Applying methods from bionics, and biomimetics, Festo's researchers and engineers studied the way kangaroos move, and applied that to the design of a robot that moves in a similar way. The robot actually saves energy from each jump and applies it to its next jump, much as a real kangaroo does.

**Tele-robots:**

- They are semi-autonomous robots controlled from a distance, chiefly using Wireless network. It is a combination of two major subfields, teleoperation and telepresence. Teleoperation indicates operation of a machine a, a distance. Telepresence refers to a set of technologies which allow a person to feel as if they were present, to give the appearance of being present.

## 3.4 INDUSTRIAL ROBOTICS

Today's industry cannot be imagined any longer without industrial robotic manipulators, which can be divided into three different groups. In the first group we classify the industrial robots which have the role of master in a robot cell. A robot cell usually compromises one or more robots, workstations, storage buffers, transport systems and numerically controlled machines. In the second group there are the robots which are slaves within the robot cell. In the third group we include the industrial robots which are used in special applications.

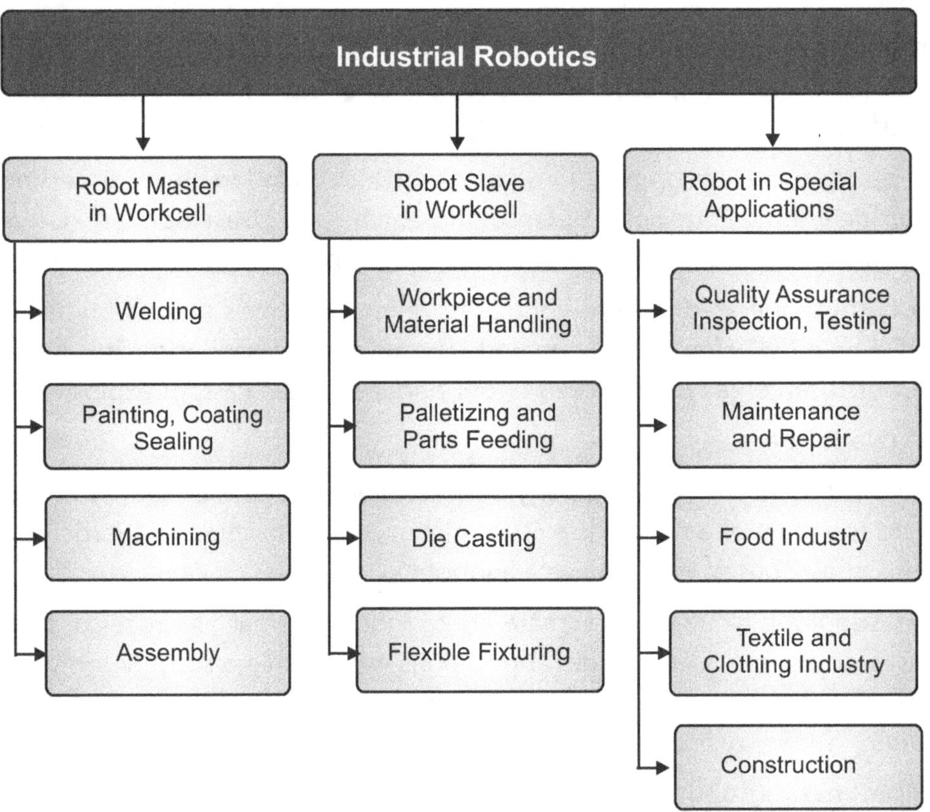


**Fig. 3.17: Industrial Robotics**

**Objectives of using Industrial Robots:**

To achieve following objectives use of industrial robots is increasing.

(a) To reduce production time.

(b) To minimise the labour requirement.

(c) To raise the quality level of products.

(d) To increase productivity.

(e) To improve existing manufacturing process.

(f) To enhance the life of production machines.

(g) To minimise the kids of man-hours on account of accidents and disease.

(h) To make the reliability and applicability of new high speed production processes and their related machinery possible.

(i) To take advantage of fair free continuous depender of robots, because the human beings are always corns to experience fatigue when put to continuous working

**Advantages:**

(a) Lifting and moving heavy objects.

(b) Working in hostile environments.

(c) Providing repeatability and consistency.

(d) Working during unfavourable hours.

(e) Performing dull or monotonous jobs.

(f) Increasing productivity, safety, efficiency and quality of products.

(g) Achieving more accuracy than human beings.

**Disadvantages:**

(a) The robots lack capability to respond in emergencies.

(b) The initial and installation costs of equipment of robots are quite high.

(c) They roads they replace human workers does causing recentment among workers.

**Common Types of Industrial Robots:**

**Articulated:** This robot design features rotary joints and can range from simple two joint structures to 10 or more joints. The arm is connected to the base with a twisting joint. The links in the arm are connected by rotary joints. Each joint is called an axis and provides an additional degree of freedom, or range of motion. Industrial robots commonly have four or six axes.

**Cartesian:** These are also called rectilinear or gantry robots. Cartesian robots have three linear joints that use the Cartesian co-ordinate system (X, Y, and Z). They also may have an attached wrist to allow for rotational movement. The three prismatic joints deliver a linear motion along the axis.

**Cylindrical:** The robot has at least one rotary joint at the base and at least one prismatic joint to connect the links. The rotary joint uses a rotational motion along the joint axis, while the prismatic joint moves in a linear motion. Cylindrical robots operate within a cylindrical-shaped work envelope.

**Polar:** Also called spherical robots, in this configuration the arm is connected to the base with a twisting joint and a combination of two rotary joints and one linear joint.  The axes form a polar co-ordinate system and create a spherical-shaped work envelope.

**SCARA:** Commonly used in assembly applications, this selectively compliant arm for robotic assembly is primarily cylindrical in design. It features two parallel joints that provide compliance in one selected plane.

**Delta:** These spider-like robots are built from jointed parallelograms connected to a common base. The parallelograms move a single EOAT in a dome-shaped work area. Heavily used in the food, pharmaceutical, and electronic industries, this robot configuration is capable of delicate, precise movement.

Typical industrial robots are articulated and feature six axes of motion (6 degrees of freedom). This design allows maximum flexibility. Six-axis robots are ideal for:

- Arc Welding.
- Spot Welding.
- Material Handling.
- Machine Tending.
- Other Applications.

## Components of Robot:

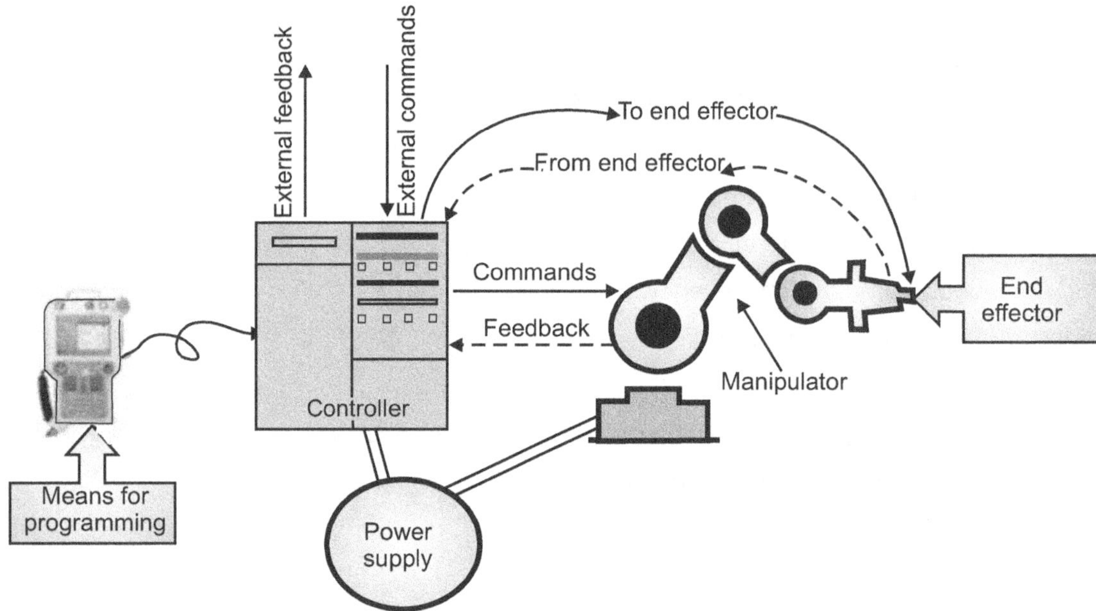


**Fig. 3.18: Components of Robots**

1. Base.

2. Manipulator arm.

3. End effector.

4. Actuators and transmission.

5. Controller.

6. Sensors.

**Base:** The base maybe fixed or mobile

**Manipulator arm :** It consists of base, arm and wrist similar to a human arm. It also includes power source either electric, hydraulic or pneumatic on receiving signals from robot controller this mechanical unit will be activated. The movement of the manipulator can be in relation to its coordinate system. Which may be Cartesian, Cylindrical etc.

Depending on the controller, the movement may be a point to point motion or continuous motion.

- The manipulator is composed of 3 divisions.

- The major linkages.

- The minor linkages (wrist components).

## The end effectors (gripper or tool)

In robotics, the physical construction of the robot consists of a body, arm, and wrist of the machine. The body is attached to the base and arm assembly attached to the body. At the end of the arm, a wrist is attached, which consists a number of components to allow variety position. Relative motion between body, arm, and wrist are provided by joints. The joint motions are either rotating or sliding. The body, arm, and wrist assembly are also called as a manipulator. Attached to the robot wrist is a hand (End effector). The end effector is not considered as part of the robot. The arm and the body joints keep the end effector in It is also called Gripper.

## Types of End-Effectors:

1. Grippers

2. Tools

**1. Grippers:** Grippers would be utilized to grasp an object, usually the work part and hold it during the robot work cycle. The types of material used for grippers depends on the orientation of part and friction between parts and grippers. The applications include material handling, machine loading, unloading, palletizing and other similar operations.

- The operations performed by grippers are.
- To move the fingers with commanded speeds.
- To gripper handled part with not more than specified force.
- To open, move and also the fingers at specified positions.
- To measure the dimension of a part.

Grippers are further classified as,

- Mechanical grippers.
- Magnetized grippers.
- Suctions or vacuum cups.
- Hooks.
- Scopes or ladles.

**Mechanical grippers:** Mechanical gripper in an end effector, that uses mechanical fingers actuated by a mechanism to grasp an object. The fingers are either attached to the mechanism.

The function of the gripper mechanism is to translate some form of power input into the grasping action.

The power input is supplied from the robot and can be pneumatic, hydraulic, and electrical.

The mechanism must be able to open and close the fingers.

**Magnetized grippers:** Magnetic grippers are be divided into 2 types:

(i) Electromagnetic.

(ii) Permanent magnetic.

**Electromagnetic:** Electromagnetic grippers include a controller unit and a DC power for handling the materials. This type of grippers is easy to control and very effective in releasing the part at the end of the operation than the permanent magnets. If the work part gripped is to be released, the polarity level is minimized by the controller unit before the electromagnet is turned-off.

**Permanent magnetic:** The permanent magnets do not require any sort of external power as like the electromagnets for handling the materials. After this gripper grasps a workpart, an additional device called as stripper push-off pin will be required to separate the workpart from the magnet. This device is incorporated at the sides of the gripper.

- This gripper only requires one surface to grasp the materials.
- The grasping of materials is done very quickly.
- It does not require separate designs for handling different size of materials.
- It is capable of grasping materials with holes, which is unfeasible in the vacuum grippers.

**Suctions or vacuum cups:** It is used to leave the flat glass of the work. This type of gripper is typically made of elastic material such as rubber, soft plastic. The shape of the vacuum is shown in figure commonly used round type. It means of removing the air between the up and part surface to create the vacuum tin required the vacuum pump and venture in of common devices and for those purposes.

The lift capacity of suction cup depends upon the effective area of the cup. The negative air pressure between the cup and the object.

$$F = P \times A$$

where,
$F$ = Force,

$P$ = Negative pressure,

$A$ = Total effective area of the suction cup used to create the vacuum.

**Hooks gripper:** Hooks can be used as end effectors to handle containers of parts to load and unload parts, hanging from, power head conveyors. The items to be handled by a hook must have some soft handle to enable the hook to hold it.

**Scopes or ladles:** Scopes or ladles can be used to handle certain materials in liquid or powder form, chemically in liquid or powder form, good materials etc. that can be handled by a robot using method of handling.

### Types of Mechanical gripper actuation:

There are various ways of classifying mechanical grippers and their actuating mechanisms. One method is according to the type of finger movement used by the gripper. In this classification, the grippers can actuate the opening and closing of the Rogers by one of the following motions:

1. Pivoting movement.                    2. Linear or translational movement.

### 1. Pivoting Movement:

In the pivoting movement. The fingers rotate about fixed pivot points on the gripper to open and close. The motion is usually accomplished by some kind of linkage mechanism.

In the linear movement, the Fingers open and close by moving in parallel to each other. This is accomplished by means of guide rails so that each finger base slides along a guide rail during actuation. The translational finger movement might also be accomplished by means of a linkage which would maintain them in a parallel orientation to each other during actuation.

Mechanical grippers can also be classed according to the type of kinematic device used to actuate the Roger movement. In this classification we have the following types:

(i) Linkage Actuation,

(ii) Gear and Rack Actuation,

(iii) Cam Actuation,

(iv) Screw Actuation,

(v) Rope and Pulley Actuation.

(i) **Linkage Actuation:** The linkage category covers a wide range of design possibilities to actuate the opening and closing of the gripper. The design of the linkage determines how the input force 'Fa' to the gripper is converted into the gripping force 'Fg' applied by the fingers. The 'linkage configuration also determines other operational features such as how wide the gripper fingers will open and how quickly the gripper will actuate.

(ii) **Gear and Rack Actuation:** These are some other mechanism that would provide linear motion, movement of the rack would drive to partial pinion gears and these would, in turn, open and close the finger.

(iii) **Cam Actuation:** The cam actuated gripper Includes a vanity possible designs, one of which is shown in Fig. cam and follower arrangement, often using and spring-loaded follower can provide the opening and closing action of the gripper.

For example, movement of the cam in one direction would force the gripper to open. while movement of the cam in the opposite direction would cause the spring to force the gripper to close. The advantage of this arrangement is that the spring action would accommodate different sized parts. This might be desirable.

For example, in a machining operation where a single gripper is used to handle the raw work part and the finished part. The finished part might be significantly smaller after machining.

**(iv) Screw Actuation:** The screw is turned by a motor. Usually accompanied by a speed reduction mechanism. When the screw is rotated in one direction, this causes a threaded block to be translated in one direction. When the screw is rotated in the opposite direction. The threaded block moves in the opposite direction. The threaded block is, in turn, connected to the gripper Buyers to cause the corresponding opening and closing action.

**(v) Rope and Pulley Actuation:** Rope and pulley actuating can be designed to open and closing a mechanical gripper.

## Construction the path in the grippers:

There are two ways of construction the path in the gripper.

(i) Physical construction of the path within the finger.

(ii) Holding the path by friction between the finger and the work path.

**(i) Physical type:** In this approach, the gripper fingers enclose the part to some extent, thereby constraining the surface of the fingers to be in the approximate shape of the part geometry.

**(ii) Holding Method:** In this approach, the fingers must apply a force that is sufficient for friction, to retain the part against gravity, acceleration and any other force that might arise during the holding portion of the work cycle.

The fingers or the pads attached to the fingers which make contact with the part. This tends to increase the coefficient of friction between the part and the contacting finger surface.

If a force of sufficient magnitude is applied against the part in a direction parallel to the friction surfaces of the fingers as shown above fig. the part might ship out of the gripper.

The following force equal can be used to determine the required magnitude of the gripper force as a function of these factors.

## Robot Controller Unit (RCU):

The instructions to the robot to perform the desired tasks are given as an input through the keyboard of this unit. The controller converts the input programs to suitable signals which activate the manipulator to perform the desired tasks.

Types of robot arms OR function line diagram representation of robot arms:

The arms of the robot classified as following:

1. **Cartesian robot:** The Cartesian robot has the simplest configuration with prismatic joints. The work envelope of a Cartesian robot is cuboidal. It has large work volume but low density. It consists of 3 linear axes.

2. **Cylinder robot:** Cylinder robot makes use of two perpendicular prismatic joint and one revolute joint. The work envelope of cylinder robot approximates to a cylinder. It consists of one linear and one rotary motion.

3. **Polar robot (Spherical):** The Polar robot consists of a rotating base, a telescopic link which can be raised or lowered about a horizontal revolute joint. It has a work envelope of a partial spherical shell. It consists of one linear and two rotary axes.

4. **Joint arm (Articulated arm):** Joint arm robot is also known as an anthropomorphic robot. It functions similar to the human arm. It consists of two straight links. Similar to the human forearm and upper arm. These two links are mounted on a rotary table and have a work envelope of spherical shape. It is the most dexterous one since all the joints are revolute joints. It consists of 3 rotary axes.

The Four Common types of arm configurations and their work envelopes.

**Degrees of Freedom:**

The number of independent motions or movement that an object can perform in its workspace is known as a Degree of Freedom of the object.

The number of independent variables required to define the position and orientation of a body in its workspace is known as Degree of Freedom of the body.

**3 DOF Motions**

A rigid body free to move on a plane has got 3-degrees of freedom. i.e., 2 translations and 1 revolution about the normal to the plane of motion. This kind of manipulator with 3 degrees of freedom is known as a planar manipulator.

**6 DOF Motions**

A rigid body free to move in space has got 6-degrees of freedom. i.e., 3 translations and 3 revolutions. A manipulator with 6 degrees of freedom is known as a spatial manipulator.

However, there are many manipulators that do not have full 6 degrees of freedom. Since not all tasks require 6 degrees of freedom.

**Calculation of Degree of Freedom for a Manipulator:**

The degree of freedom of a manipulator is the number of independent motions possessed by that manipulator in its working space. The independent motions depend on the number and type of links and joints in that manipulator. The Degree of Freedom of the joints decides the DoF of the manipulator.

The relation between the Number of Degree of Freedom can be given by the Kutzbach Criterion. And is as follows:

Where,

$N$ = No. of Individual Motions in the Manipulator or Degrees of Freedom.

$m$ = Maximum number of Degrees of Freedom of a free body in the Workspace.

3 for Planar Manipulator.

6 for Spatial Manipulator.

$n$ = Total number of links in the Manipulator.

$j_i$ = Number of joints having 'i' number of DoF restrictions or having $(m - i)$ DoF.

The DOF is equal to the number of links in an open kinematic chain. Robot with 6 DOF is called as a spatial manipulator (arm, body and wrist). Robot is more than 6 DOF is called as a redundant manipulator.

## 3.5 4D PRINTING TECHNOLOGY (4-DIMENSIONAL PRINTING)

4-dimensional printing (4D printing; also known as 4D bio-printing, active origami, or shape-morphing systems) uses the same techniques of 3D printing through computer-programmed deposition of material in successive layers to create a three-dimensional object. However, 4D printing adds the dimension of transformation over time. It is therefore a type of programmable matter, wherein after the fabrication process, the printed product reacts with parameters within the environment (humidity, temperature etc.,) and changes its form accordingly. The ability to do so arises from the near infinite configurations at a micrometer resolution, creating solids with engineered molecular spatial distributions and thus allowing unprecedented multifunctional performance.

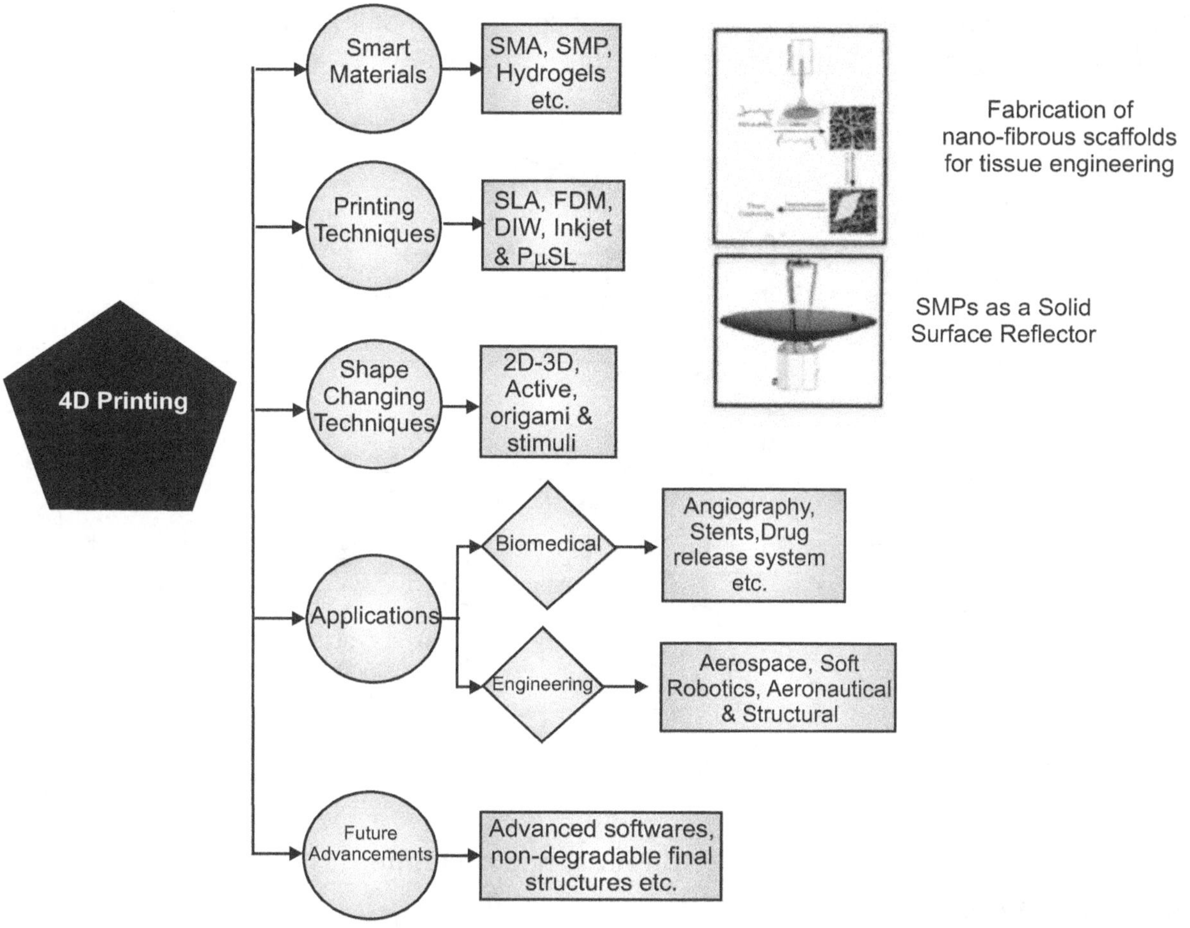



**Fig. 3.19: 4D Printing Technology**

- The ability to program physical and biological materials to change shape, change properties, and compute outside of silicon based matter."

- Goal: Take nano/biomolecular self-assembly and apply to human scale in order to build infrastructure more efficiently.

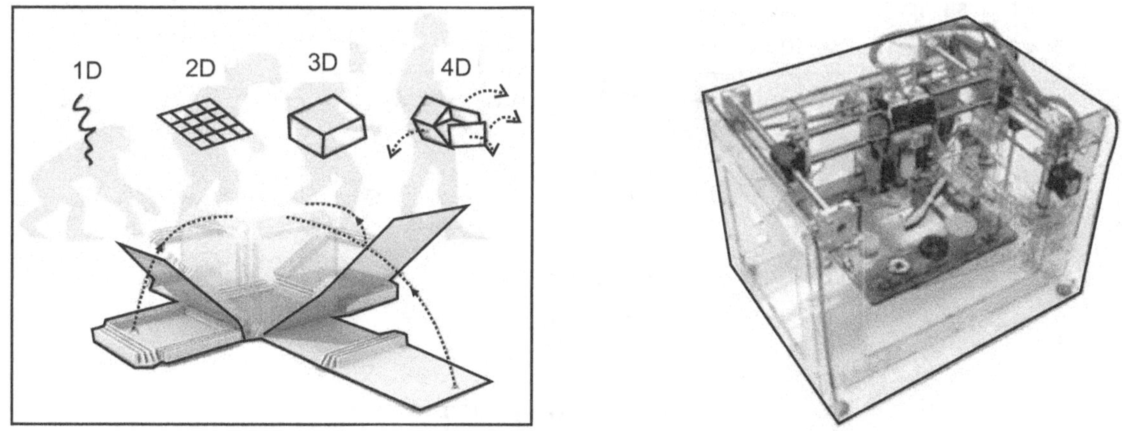



**Fig. 3.20: 4D Printer**

4D printing technology drives a significant transformation in the medical field. It has broader application in tissue engineering, chemotherapy and self-assembling human scale biomaterials. Now there is a development from 3D medical devices to 4D medical devices such as manufacturing of biomedical splint, stents, bio-printing and orthodontic devices. It manufactures implant as planned which, grows with human growth. 4D printing handles such difficult situation easily with better accuracy. By the application of this technology, there are endless possibilities which potentially impact all sectors like medical, manufacturing and education. In the medical field, there is an excellent contribution of 4D printing technology to minimize the procedure of surgery. For patient's body, doctors can use self-deforming component to treat any abnormality.

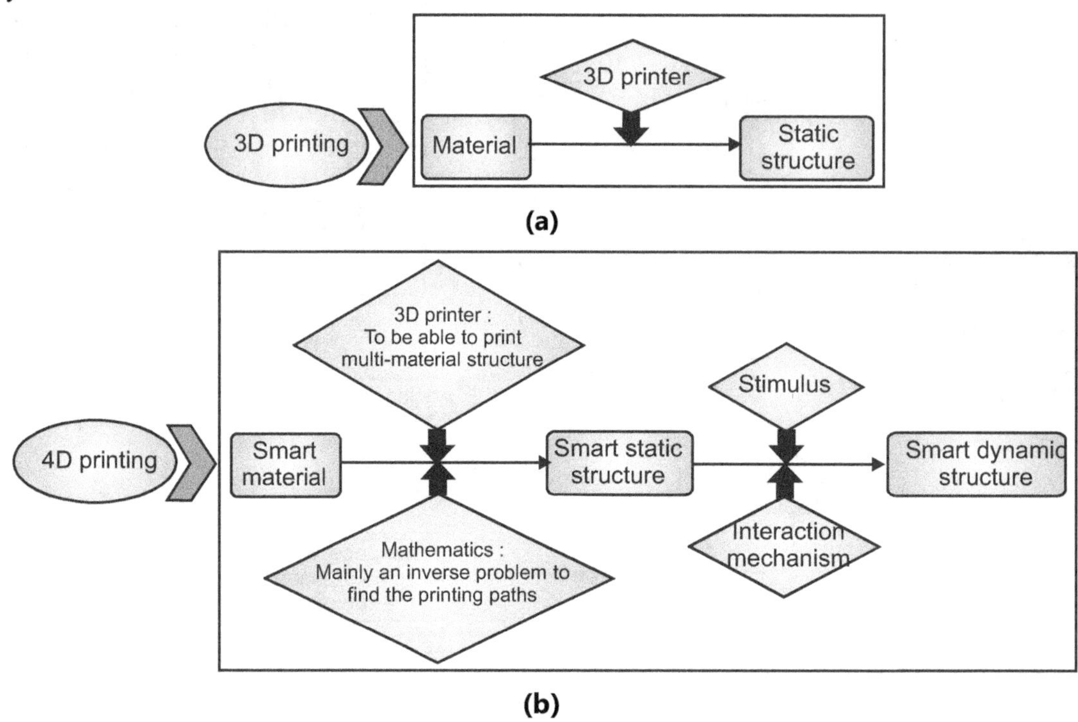


**Fig. 3.21: 3D Printer and 4D Printer**

## 4D Printing Techniques:

### Fiber Architecture:

One of the composite polymers that Tibbets et. al. printed, reacting when submerged underwater. Most 4D printing systems utilize a network of fibers that vary in size and material properties. 4D printed components can be designed on the macro scale as well as the micro scale. Micro scale design is achieved through complex molecular/fiber simulations that approximate the aggregated material properties of all the materials used in the sample. The size, shape, modulus, and connection pattern of these material building blocks have a direct relationship to the deformation shape under stimulus activation.

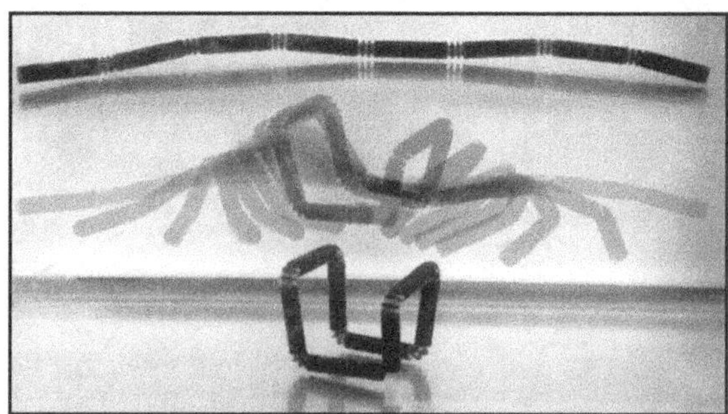

**Fig. 3.22: 4D Printing Technology**

### Hydro-reactive polymers/hydrogels:

Skylar Tibbits is the director of the Self-Assembly Lab at MIT, and worked with the Stratasys Materials Group to produce a composite polymer composed of highly hydrophilic elements and non-active, rigid elements. The unique properties of these two disparate elements allowed up to 150% swelling of certain parts of the printed chain in water, while the rigid elements set structure and angle constraints for the transformed chain. Tibbits et al. produced a chain that would spell "MIT" when submerged in water, and another chain that would morph into a wireframe cube when subjected to the same conditions.

## Cellulose Composites:

Thiele et al. explored the possibilities of a cellulose-based material that could be responsive to humidity. They developed a bilayer film using cellulose steraroyl esters with different substitution degrees on either side. One ester had a substitution degree of 0.3 (highly hydrophilic) and the other had a substitution degree of 3 (highly hydrophobic.) When the sample was cooled from 50 °C to 22 °C, and the relative humidity increased from 5.9% to 35%, the hydrophobic side contracted and the hydrophilic side swelled, causing the sample to roll up tightly. This process is reversible, as reverting the temperature and humidity changes caused the sample to unroll again.

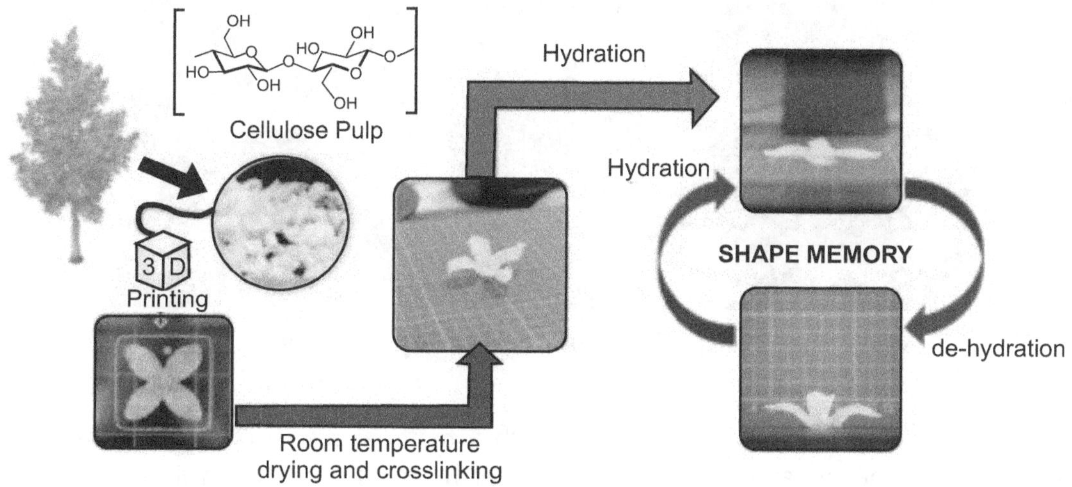


**Fig. 3.23: Cellulose Composites**

## Thermo-reactive polymers/hydrogels:

Poly (N-isopropylacrylamide), or pNIPAM is a commonly used thermo-responsive material. A hydrogel of pNIPAM becomes hydrophilic and swollen in an aqueous solution of 32 °C, its low critical solution temperature. Temperatures above that start to dehydrate the hydrogel and cause it shrink, thus achieving shape transformation. Hydrogels composed of pNIPAM and some other polymer, such as 4-hydroxybutyl acrylate (4HBA,) exhibit strong reversibility, where even after 10 cycles of shape change there is no shape deformation. Shannon E. Bakarich et al. created a new type of 4D-printing ink composed of ionic covalent entanglement hydrogels that have a similar structure to standard double-network hydrogels. The first polymer network is cross-linked with metal cations, while the second is cross-linked with covalent bonds. This hydrogel is then paired with a pNIPAM network for toughening and thermal actuation. In lab testing, this gel showed a shape recovery of 41%-49% when the temperature increased 20–60 °C (68–140 °F), and then was restored to 20 °C. A fluid controlling smart valve printed from this material was designed to close when touching hot water and open when touching cold water. The valve successfully stayed open in cold water and reduced the flow rate of hot water by 99%. This new type of 4D-printed hydrogel is more mechanically robust than other thermally actuating hydrogels and shows potential in applications such as self-assembling structures, medical technology, soft robotics, and sensor technology.

## Digital Shape-Memory Polymers:

Shape-memory polymers (SMPs) are able to recover their original shape from a deformed shape under certain circumstances, such as when exposed to a temperature for a period of time. Depending on the polymer, there may be a variety of configurations that the material may take in a number of temperature conditions. Digital SMPs utilize 3D-printing technology to precisely engineer the placement, geometry, and mixing and curing ratios of SMPs with differing properties, such as glass transition or crystal-melt transition temperatures. Yiqi Mao et al. used this to create a series of digital SMP hinges that have differing prescribed thermo-mechanical and shape memory behaviors, which are grafted onto rigid, non-active materials. Thus, the team was able to develop a self-folding sample that could fold without interfering with itself, and even interlock to create a more robust structure. One of the projects include a self-folding box modelled after a USPS mailbox.

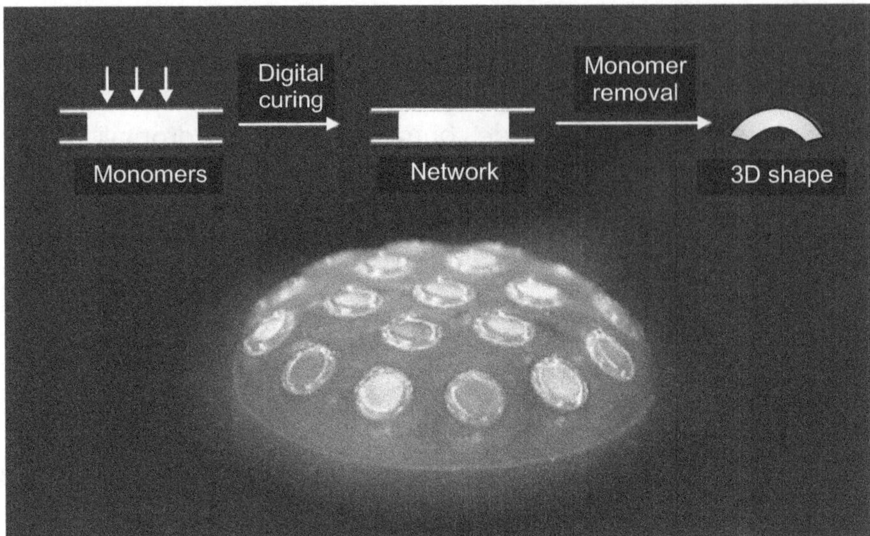


**Fig. 3.24: Digital Shape-Memory Polymers**

**Stress relaxation:**

Stress relaxation in 4D printing is a process in which a material assembly is created under stress that becomes "stored" within the material. This stress can later be released, causing an overall material shape change.

**Thermal photo-reactive polymers:**

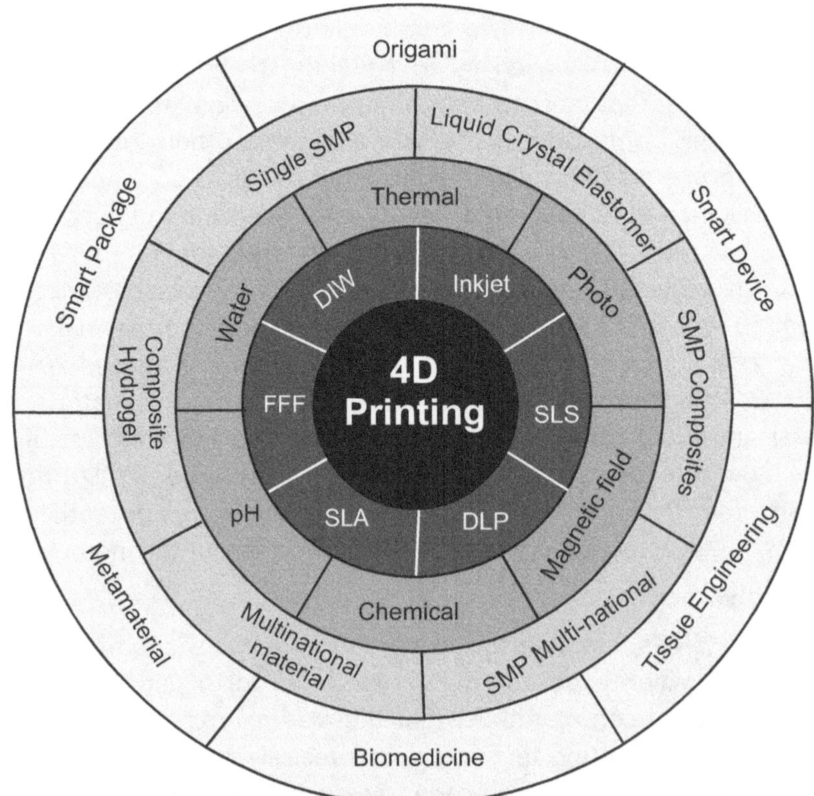


**Fig. 3.25: 4D Printer Wheel**

This technology takes advantage of temperature driven polymer bending by exposing the desired bending seams to focused strips of intense light. These bending seams are printed in a state of stress but do not deform until exposed to light. The active agent that induces bending in the material is heat transmitted by intense light. The material itself is made of chemical photo-reactive polymers. These compounds use a polymer mixture combined with a photoinitiator to create an amorphous, covalently cross-linked polymer. This material is formed into sheets and loaded in tension perpendicular to the desired bending crease. The material is then exposed to a specific wavelength of light, as the photoinitiator is consumed it polymerizes the remaining mixture, inducing photoinitiated stress relaxation. The portion of material exposed to the light can be controlled with stencils to create specific bending patterns. It is also possible to run multiple iterations of this process using the same material sample with different loading conditions or stencil masks for each iteration. The final form will depend on the order and resulting form of each iteration.

## Application of 4D Printing:

### Electrical and Magnetic Smart Materials:

The electrical responsive materials that exist today change their size and shape depending on the intensity and/or direction of an external electric field. Polyaniline and polypyrrole (PPy) are, in particular, good conducting materials and can be doped with tetrafluoroborate to contract and expand under an electric stimulus. A robot made of these materials was made to move using an electric pulse of 3V for 5 seconds, causing one leg to extend, then removing the stimulus for 10 seconds, causing the other leg to move forward. Research on carbon nanotubes, which are biocompatible and highly conductive, indicates that a composite made of carbon nanotube and a shape memory specimen has a higher electrical conductivity and speed of electro-active response than either specimen alone. Magnetically responsive ferrogels contract in the presence of a strong magnetic field and thus have applications in drug and cell delivery. The combination of carbon nanotubes and magnetically responsive particles has been bioprinted for use in promoting cell growth and adhesion, while still maintaining a strong conductivity.

### Commerce and Transportation:

Skylar Tibbits elaborates on future applications of 4D-printed materials as programmable products that can be tailored to specific environments and respond to factors such as the temperature, humidity, pressure, and sound of one's body or environment. Tibbits also mentions the advantage of 4D-printing for shipping applications - it will allow products to be packaged flat to later have their designed shape activated on site by a simple stimulus. There is also the possibility of 4D-printed shipping containers that react to forces in transit to uniformly distribute loads. It is very likely that 4D-printed materials will be able to repair themselves after failure. These materials will be able to self-disassemble, making their constituent parts easy to recycle.

### Medical Field:

**Table 3.1: 4D printing applications in the medical field**

| Sr. No. | Medical Application | Description |
|---|---|---|
| 1. | Smart sense | <ul><li>4D printing could develop stents which can get expanded and take required shape with the help of heat of the patient body.</li><li>This latest solution act quickly to save a patients life for complicated surgery.</li><li>A shape gets changed w.r.t. time and temperature.</li></ul> |
| 2. | Organ printing | <ul><li>This emerging technology applies to fabricate complex 3D organ.</li><li>Used to print organs where own cells of the patient are applied and save a life.</li><li>It is a promising solution for organ shortages.</li></ul> |

| | | |
|---|---|---|
| 3. | Smart multi-material printing | • Use UV curable polymer in layer by layer technique.<br>• It is a new process for printing customizable smart multi-material printing of medical implants.<br>• Clearly, show multi parts of the body in the 3D printed body. |
| 4. | Dyspnen (Breathing problem) | • 4D printing technology save lives of babies which are suffering from Dyspnen (breathing problems).<br>• It quickly creates a medical implant which on change shape over time as babies grew and helped them keep breathing. |
| 5. | Smart medical implants and tissue engineering | • 4D printing centres exciting possibilities of the shape-shifting material.<br>• Its potential application could use for medical implants and tissue engineering that change their shape inside the body.<br>• Used for regeneration of those tissues where mechanical properties vary dynamically as our body gets active, such as muscle, bone, and cardiovascular tissues. |
| 6. | Printing of heart, kidney and liver | • In the future, 4D printing can print heart, kidney and liver by using the smart material.<br>• Ability to print these parts which have great flexibility with precise fit and perfect genetic match. |
| 7. | Skin Graft | • It has great possibilities to print skin graft like original colour of the patients.<br>• Also useful for the burn patient which will easily implant in the patient body and grows like an original. |
| 8. | Manufacturing of smart medical devices | • This technology can manufacture 3D-printed complex smart medical devices which have excellent functional properties.<br>• Adjusted as per requirement of surgery concerning time. |
| 9. | Complex surgery | • It could produce a haptic model in represent the movement of the body as well as appearance.<br>• In future, it can well be adopted for quite complicated surgery which is not possible by other manufacturing technologies.<br>• It can produce a model by using a smart material with the help of CT and MRI scan which accurately replicate hand or another body movement.<br>• It can represent anatomical details in a precise and accurate manner. |

## Benefits of 4D printing:

4D printing technology can develop products which can change characteristics and properties from environment change like temperature with time. Various benefits of 4D printing are as follows:

- **Capabilities of the printed smart products:** 4D printing provides a broad capability for users to print products with smart material which has broad applications in engineering, medical, dentistry and material science.

- **Change of product shape as per the requirement:** Changes in the product shape of a 3D printed model related as a function of time where printed product reacts with the parameter like temperature, humidity etc. to change its form as per the requirement.

- **Create innovation:** This technology innovates product during the design and development stage. The technology is also successfully used for research.

- **Self-assembly:** This technology tends to produce a product which has self-assemble material property due to the usage of smart material.

## 3D Printing Technology:

The 3D printing process builds a three-dimensional object from a computer-aided design (CAD) model, usually by successively adding material layer by layer, which is why it is also called additive manufacturing, unlike conventional machining, casting and forging processes, where material is removed from a stock item (subtractive manufacturing) or poured into a mold and shaped by means of dies, presses and hammers.

- 3D printing or additive manufacturing is a process of making three dimensional solid objects from a digital file.

- The creation of a 3D printed object is achieved using additive processes. In an additive process an object is created by laying down successive layers of material until the object is created. Each of these layers can be seen as a thinly sliced horizontal cross-section of the eventual object.

- 3D printing is the opposite of subtractive manufacturing which is cutting out/hollowing out a piece of metal or plastic with for instance a milling machine.

- 3D printing enables you to produce complex shapes using less material than traditional manufacturing methods.

## 3D Scanning:

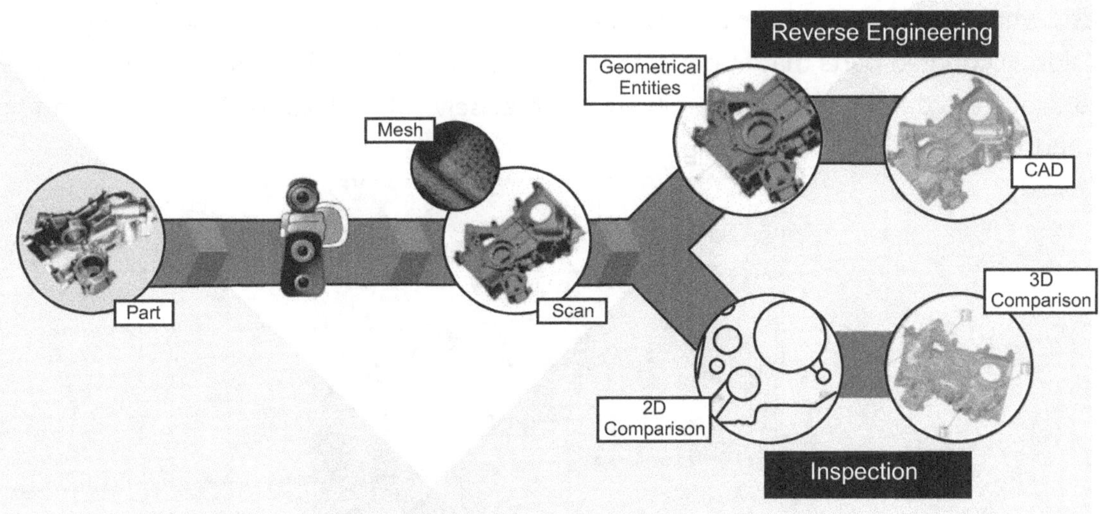


**Fig. 3.26: 3D Scanning Process**

3D scanners are tri-dimensional measurement devices used to capture real-world objects or environments so that they can be remodeled or analyzed in the digital world. The latest generations of 3D scanners do not require contact with the physical object being captured.

3D scanners can be used to get complete or partial 3D measurements of any physical object. The majority of these devices generate points or measures of extremely high density when compared to traditional "point-by-point" measurement devices.

**Type 3D Scanning:**

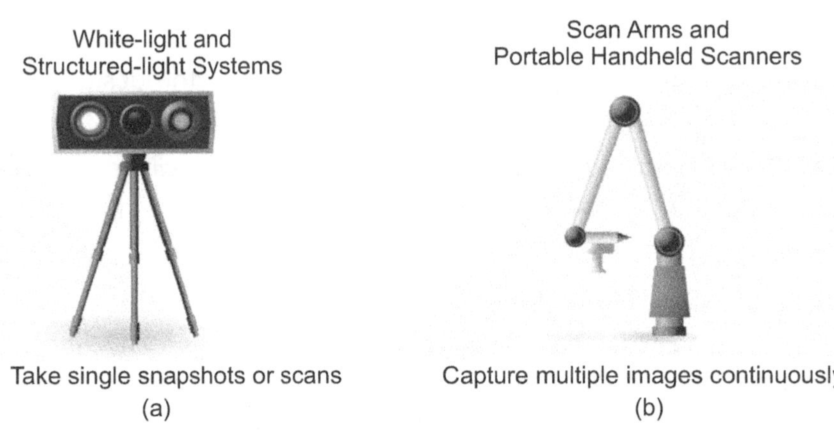


**Fig. 3.27: Types of 3D Scanning**

**3-D Scanning with Reverse Engineering:**

Reverse engineering involves capturing the technological details and composition of an object in order to augment, improve or recreate it. Used in a variety of industries, such as industrial engineering, automotive, aviation, manufacturing, electronics and many others, it allows replicating the initial object by capturing, analyzing and disassembling its digital 3D replica. Reverse engineering is paramount to multiple industries such as manufacturing, automotive, aerospace, and in a wider sense, even areas such as healthcare. It has a broad range of applications, including product improvement and reconstruction, design augmentation and many others. The world today would not be the same without it.

3D scanners measure complex objects very quickly, and can speed up your design workflow tremendously when real-life references are involved. With the ability to capture and modify physical shapes, you can design 3D printed parts that fit perfectly on existing products of all kinds. 3D printed jigs allow you to repeatedly locate a drill or saw, or assemble parts precisely with adhesive. Create close-fitting, reusable masks for sandblasting, painting, or etching.

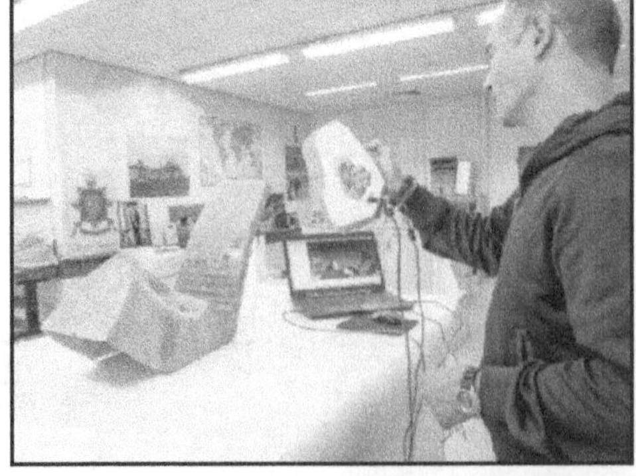

**Fig. 3.28: 3D Scanning with Reverse Engineering**

## Applications of 3D printing:

- Construction industry and civil engineering
- Design process
- Entertainment
- 3D photography
- Law enforcement
- Reverse engineering
- Real estate
- Virtual/remote tourism
- Cultural heritage
  - Michelangelo
  - Monticello
  - Cuneiform tablets
  - Kasubi Tombs
  - "Plastico di Roma antica"
  - Other projects
- Medical CAD/CAM
- Quality assurance and industrial metrology
- Circumvention of shipping costs and international import/export tariffs

### Table 3.2: Comparative analysis of 4D printing with the 3D printing

| Sr. No. | Characteristics | 3D printing | 4D printing |
|---|---|---|---|
| 1. | Built Process | • 3D printing repeats a 20 structure layer by layer from bottom to top. | • 4D printing is the extension of 3D printing. |
| 2. | Material Used | • Uses thermoplastics, ceramics, metals, biomaterials or nanomaterials. | • Uses smart, multi-material and self-assembling material to build as object, which change its shape after getting manufactured.<br>• There is a need to develop new materials as per the requirements of the applications. |
| 3. | Flexibility | • No flexibility, characterized by rigidity. | • Having flexibility, characterized by flexibility. |
| 4. | Object shape flexibility | • No flexibility, characterized by rigidity. | • Object shape is changed over time and with the change in temperature. |
| 5. | Programming of material | • Not use any programmable and advanced material. | • Use programmable and advanced (most new) material that can provide various functionalities. |
| 6. | Applications | • Its application is in medical, engineering, dentistry, automobile, jewelry, toys, fashion, entertainment, aerospace, defence etc. | • Dynamically changing configuration for all applications by 3D printing. |

## Multiple Choice Questions

1. Robot is derived from Czech word ......

   (a) Rabota         (b) Robota

   (c) Rebota         (d) Ribota

2. A Robot is a ......

   (a) Programmable      (b) Multi functional manipulator

   (c) Both (a) and (b)     (d) None of the above

3. The main objective(s) of Industrial robot is to ......

   (a) To minimise the labour requirement.

   (b) To increase productivity.

   (c) To enhance the life of production machines.

   (d) All of the above.

4. The following is true for a Robot and NC Machine ......

   (a) Similar power drive technology is used in both.

   (b) Different feedback systems are used in both.

   (c) Programming is same for both.

   (d) All of the above.

5. Match the following :

   |  | Robot part | | Function |
   |---|---|---|---|
   | a. | Manipulator arm | 1. | For holding a piece or tool |
   | b. | Controllers | 2. | Move the manipulator arm and end effector |
   | c. | Drives | 3. | Number of degrees of freedom of movement |
   | d. | Gripper | 4. | Delivers commands to the actuators |

   (a) a-1, b-4, c-2, d-3     (b) a-3, b-4, c-2, d-1

   (c) a-3, b-2, c-4, d-1     (d) a-4, b-3, c-2, d-1

6. Drives are also known as ......

   (a) Actuators       (b) Controller

   (c) Sensors        (d) Manipulator

7. Clockwise of Anti clockwise rotation about the vertical axis to the perpendicular arm is provided through ......

   (a) Shoulder swivel     (b) Elbow extension

   (c) Arm sweep       (d) Wrist bend

8. Radial movement (in and out) to the manipulator arm is provided by ......

   (a) Elbow extension     (b) Wrist bend

   (c) Wrist swivel      (d) Wrist yaw

9. Industrial Robots are generrally designed to carry which of the following co-ordinate system(s).

   (a) Cartesian co-ordinate systems   (b) Polar co-ordinate systems

   (c) Cylindrical co-ordinate system   (d) All of the above

10. The Robot designed with Cartesian co-ordinate systems has ......

    (a)  Three linear movements

    (b)  Three rotational movements

    (c)  Two linear and one rotational movement

    (d)  Two rotational and one linear movement

11. The Robot designed with Polar co-ordinate systems has ......

    (a)  Three linear movements

    (b)  Three rotational movements

    (c)  Two linear and one rotational movement

    (d)  Two rotational and one linear movement

12. The Robot designed with cylindrical co-ordinate systems has ......

    (a)  Three linear movements

    (b)  Three rotational movements

    (c)  Two linear and one rotational movement

    (d)  Two rotational and one linear movement

13. Which of the following work is done by General purpose robot?

    (a)  Part picking          (b)  Welding          (c)  Spray painting          (d)  All of the above

14. The following drive is used for lighter class of Robot.

    (a)  Pneumatic drive          (b)  Hydraulic drive          (c)  Electric drive          (d)  All of the above

15. Internal state sensors are used for measuring ...... of the end effector.

    (a)  Position

    (b)  Position amd Velocity

    (c)  Velocity and Acceleration

    (d)  Position, Velocity and Acceleration

16. Which of the following sensors determines the relationship of the robot and its environment and the objects handled by it ?

    (a)  Internal State sensors

    (b)  External State sensors

    (c)  Both (a) and (b)

    (d)  None of the above

17. Which of the following is not a programming language for computer controlled robot?

    (a)  AMU          (b)  VAL          (c)  RAIL          (d)  HELP

18. In which of the following operations Continuous Path System is used

    (a)  Pick and Place

    (b)  Loading and Unloading

    (c)  Continuous welding

    (d)  All of the above

19. Flexible manufacturing systems (FMS) are reported to have a number of benefits. Which is NOT a reported benefit of FMS?

    (a)  Lead time and throughput time reduction.

    (b)  More flexible than the manufacturing systems they replace.

    (c)  Increased quality.

    (d)  Increased utilization.

20. Which of the following is NOT a 3-D printing approach?

    (a)  Computer numerical controlled (CNC) machining.

    (b)  Fused deposition modelling (FDM).

    (c)  Photopolymerization.

    (d)  None of the above.

## Answers

| Question No. | 1 | 2 | 3 | 4 | 5 | 6 | 7 | 8 | 9 | 10 |
|---|---|---|---|---|---|---|---|---|---|---|
| Answer | (b) | (c) | (d) | (a) | (b) | (a) | (c) | (a) | (d) | (a) |
| Question No. | 11 | 12 | 13 | 14 | 15 | 16 | 17 | 18 | 19 | 20 |
| Answer | (d) | (c) | (d) | (a) | (d) | (c) | (a) | (c) | (d) | (a) |

✍ ✍ ✍

# Chapter 4...

# ENERGY MONITORING, MANAGEMENT AND AUDIT



## Syllabus

4.1 **List of Bureau of Energy Efficiency Standards (BEE):** Standards and labeling Standards (HVAC)

4.2 **Energy Monitoring and Targeting:** Defining monitoring and targeting, Elements of monitoring and targeting, Methods Monitoring and Targeting System

4.3 **Energy Management and Audit:** Definition, Energy audit- need, Types of energy audit, Energy management (Audit) approach-understanding energy costs.

## About this Chapter

At the end of this chapter, students will be able to:

- List different Bureau of Energy Efficiency (BEE) standards.
- Describe methods of Energy Monitoring of Targeting.
- Identify steps for conducting Energy Audit.

## 4.1 BUREAU OF ENERGY EFFICIENCY (BEE)

- Established in 2002, under the Energy Conservation Act, 2001.
- Improve energy efficiency through various regulatory and  promotional instruments.
- Plan, manage and implement provisions under the EC Act.
    - Appliance standards and labelling.
    - Industrial energy benchmarks.
    - Energy Conservation Building Codes.
    - Monitor energy use in high energy-consumption units.
    - Certify and accredit energy auditors and energy managers.
- Provide a policy framework and direction to national energy conservation activities.
- Disseminate information and knowledge, and facilitate pilot and demonstration projects.

**Functions of the Bureau are as follows:**

- The process and energy consumption standards required to be notified;
- The labelling of certain equipment requiring some input of energy, along with the prescription about the display of standards upon such labels;
- To notify users or class of users as 'designated consumers' under this law.

**BEE created as a nodal statutory body to improve energy efficiency through:**

- Standards and Labelling for equipments/appliances (S&L).
- Energy Conservation Building Codes (ECBC).

- Energy consumption norms for Designated Consumers.
- Certification and accreditation of energy auditors and energy managers.
- Dissemination of information and best practices.
- Capacity Building.
- Establish Energy Efficiency delivery systems through Public-Private.

## Partnerships

The Act creates the Bureau of Energy Efficiency (BEE) in the center, and State Designated Agencies (SDAs) in the states.

## Bureau of Indian Standards (BIS)

Formulation & Implementation of National Standards

Production certification, Quality system certification, EMS certification etc.

**Fig. 4.1 (a)**

## Bureau of Energy Efficiency (BEE)

BEE is established to implement and monitor the Energy Conservation Act, 2001.

One of the key thrust areas of EC Act, 2001 is Standards and Labelling Programme.

**Fig. 4.1 (b)**

## Standards and Labelling Standards:

- Wide variation in energy consumption by products of manufacturers is observed.
- Information on energy consumption is often not easily available, sufficient or easy to understand from the nameplate.
- Lead to continued manufacture and purchase of inefficient equipments and appliances.
- The first mandatory minimum energy-efficiency standard was introduced in Poland during 1962.
- Russia introduced the efficiency information labels and performance standards from 1960 onwards.
- French government introduced standards for refrigerator (1968) and for freezers (1978).
- The state of California, U.S introduced the energy-efficiency standards in 1976
- Around the world 43 governments have introduced the standards and labelling during 2000 and it is increased to 65 in the year 2007.
- Recently, a number of countries have initiated programs of voluntary endorsement labelling for energy efficient products.
- Many other countries including Australia, Canada, China, Brazil, Thailand, Japan, and the United Kingdom (U.K.) have subsequently implemented national programs.

## Standards and Labelling (HVAC)

## Standard:

- Prescribe limits on the energy consumption (or minimum levels of the energy efficiency) of manufactured products.

- "Standards" commonly encompasses two possible meanings:
  - o well-defined test protocols to obtain a sufficiently accurate estimate.
  - o target limits on energy performance.
- Energy-efficiency standards are procedures and regulations that prescribe the energy performance of manufactured products, sometimes prohibiting the sale of products that are less energy efficient than the minimum standard, often called Minimum Energy Performance Standards (MEPS).

## Roles of Energy Efficiency Standard

- Provide a mechanism to exclude low energy-efficiency products from the market, and for implementing government supervision.
- Provide technical basis for establishing energy efficiency labelling programs.
- Provide technical supports for developing relative energy efficiency policies and initiatives.
- Provide technical basis for bulk procurement program.
- Provide technical basis for international harmonization.

## Label:

- "Labels" mainly give consumers the necessary information to make informed purchase.
- Energy-efficiency labels are informative labels affixed to manufactured products to describe the product's energy performance (usually in the form of energy use, efficiency, or energy cost); these labels give consumers the data necessary to make informed purchases.
- There are two types of labels:

| Comparative Label | Endorsement Label |
|---|---|
| Allow consumers to compare the energy consumption of similar products, and factor lifetime running cost into their purchasing decision. | Provide a 'certification' to inform prospective purchasers that the product is highly energy efficient for its category. |

## The Category of Energy Efficiency Comparative Label

- Comparative labels are divided:
  - o Categorical label;
  - o Continuous-scale label;
  - o Information-only label.
- Comparative label can be:
- Mandatory;
- Voluntary.

## Energy Efficiency Ratio (EER):

Ratio of the total annual amount of heat that the equipment can remove from the indoor air to the total annual amount of energy consumed by the equipment.

or

The Energy Efficiency Ratio (EER) of an HVAC cooling device is the ratio of output cooling energy (in BTU) to input electrical energy (in watts) at a given operating point.

**Table 4.1**

| Star rating (29.06.205 to 31.12.2017) | | | Star rating (01.01.2018 to 31.12.2019) | | |
|---|---|---|---|---|---|
| **Star Rating** | **Minimum EER** | **Maximum EER** | **Star Rating** | **Minimum EER** | **Maximum EER** |
| 1-Star | 3.10 | 3.29 | 1-Star | 3.10 | 3.29 |
| 2-Star | 3.30 | 3.49 | 2-Star | 3.30 | 3.49 |
| 3-Star | 3.50 | 3.99 | 3-Star | 3.50 | 3.99 |
| 4-Star | 4.00 | 4.49 | 4-Star | 4.00 | 4.49 |
| 5-Star | 4.50 | | 5-Star | 4.50 | |

**Type of Labels:**

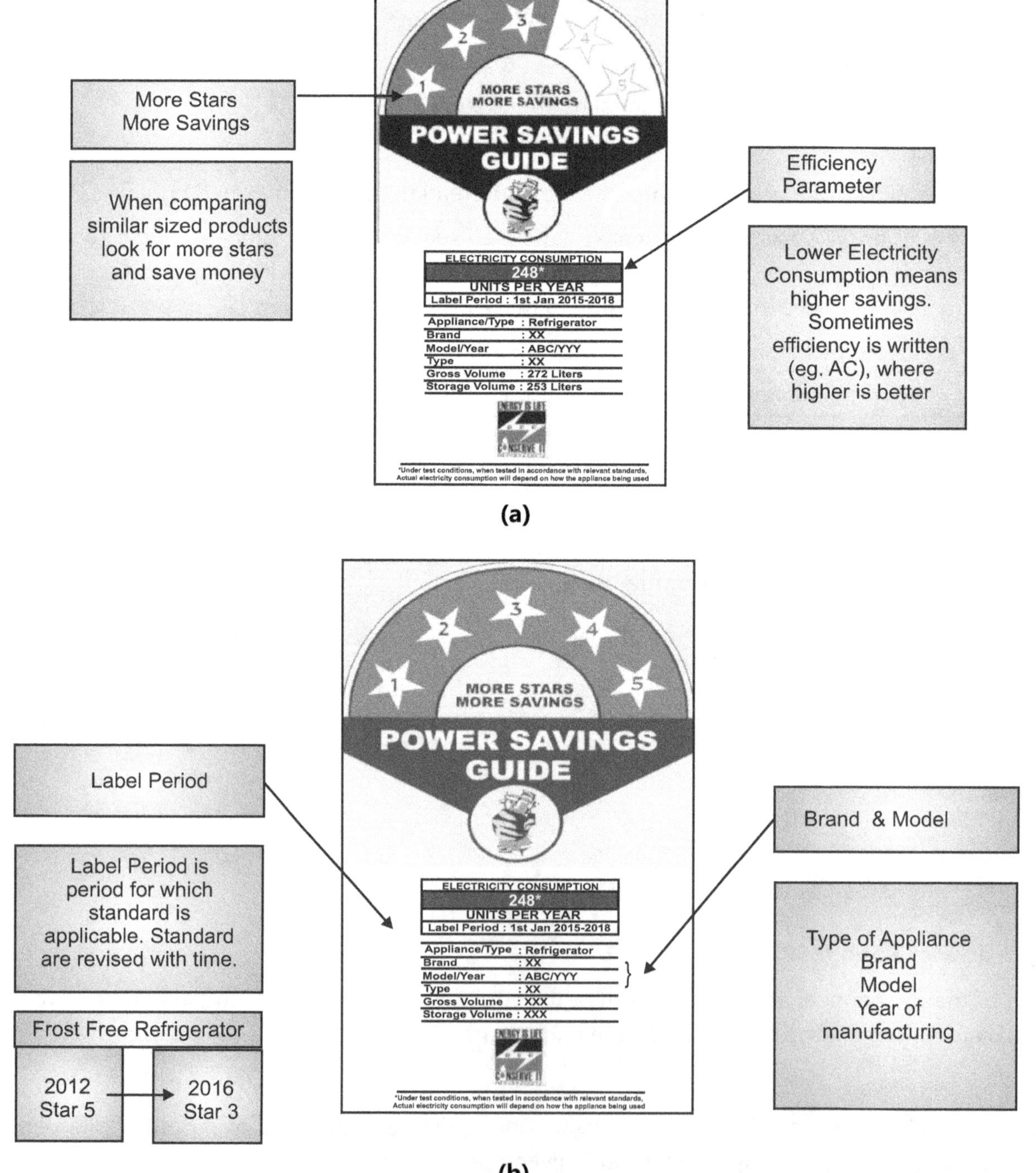


(a)

(b)

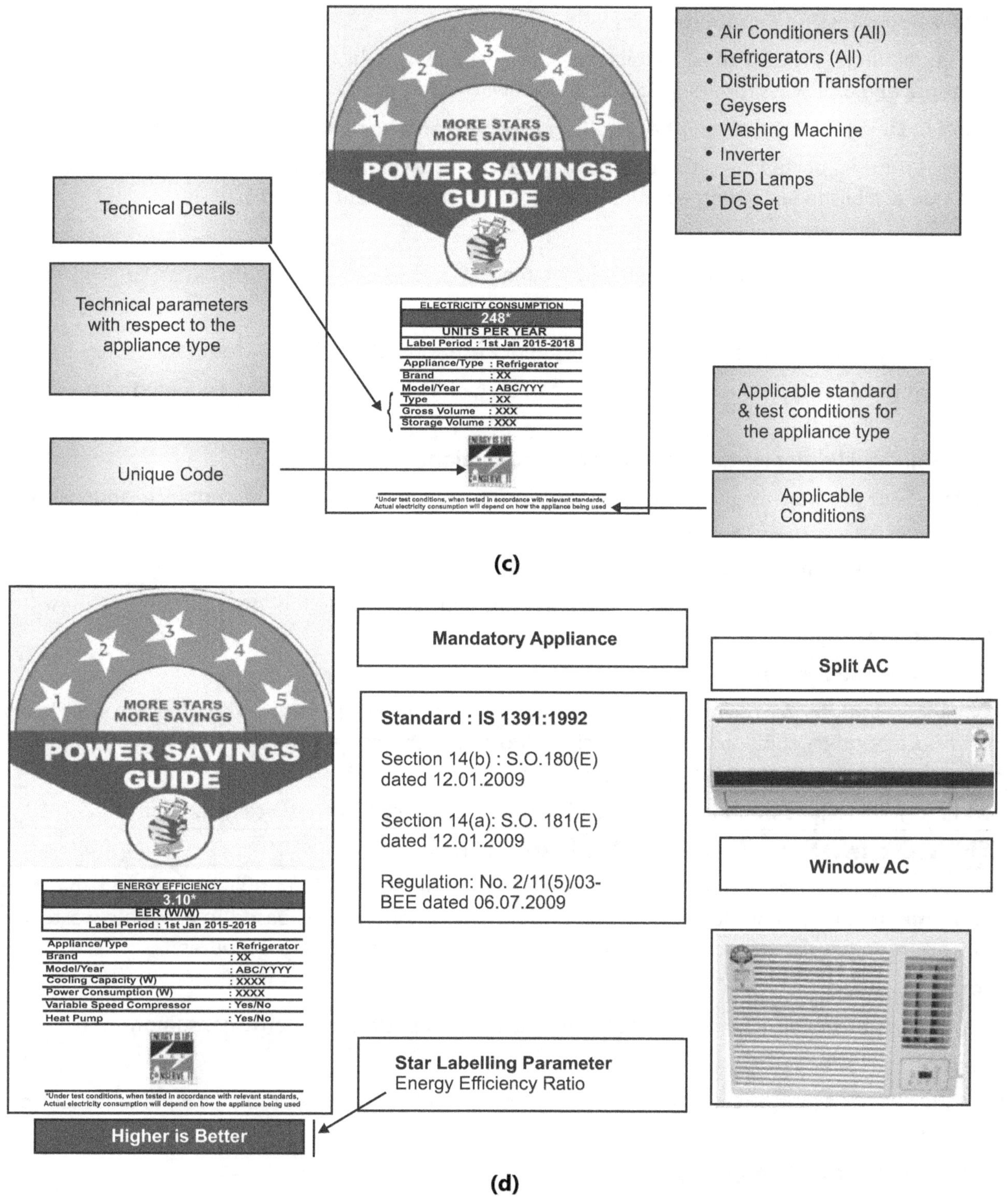


**(c)**



**(d)**

**Fig. 4.2**

## Standards and Labelling Methodology:

### Step 1: Decide whether and how to implement Energy Efficiency Labels and Standards.

- The potential impact of the standards by quantifying their predicted environmental and monetary benefits shall be addressed.
- Screening and selecting which types of products are the highest priorities.
- Assessment of the data needs for the program.
- Backing-up with the test procedures and testing facilities adopted in other countries.

### Step 2: Develop a testing Capability.

All manufacturers' product must be evaluated in a same way which requires a standard testing facility, test procedure and a process for assuring compliance with testing requirements.

### Step 3 and 4: Design and implement a labelling program and analyze and set Standards.

From consumers' perspective, the energy label is the most important element of the program. Label design can be established involving consumer research as an important element.

A Standard can be set to:

- Eliminate inefficient models currently in the market.
- Avoid import of inefficient products.
- Encourage local manufacturers to develop more economically efficient products.

Several types of analyses such as technical, market, national impact etc. are conducted to ensure that a standard achieve its purpose.

### Step 5: Design and implement a communication campaign.

Effective S&L program require a communication campaign to support acceptance and use of new standards and labels.

### Step 6: Ensure program integrity.

After the program initiation, a verification regime (to determine the product energy performance compliance to the program) is needed to ensure program's integrity.

### Step 7: Evaluate the program.

To maintain the program over the long run, the government shall monitor the program's performance together information to guide adaptation to changing circumstances and to clearly demonstrate the public that the expected benefits are actually being achieved.

Good program require periodic revision and update. Review cycle can typically range from 2 to 12 years.

## 4.2 ENERGY MONITORING AND TARGETING

Energy monitoring and targeting is primarily a management technique that is builds on the principle **"you can't manage what you don't measure"**. This energy information is based to eliminate waste, reduce and control current level of energy use and improve the existing operating procedures.

### Monitoring:

Monitoring is essentially aimed at establishing the existing pattern of energy consumption

### Targeting:

Targeting is the identification of energy consumption level which is desirable as a management goal to work towards energy conservation.

### Benefits Monitoring and Targeting:

- Increase in energy performance leads to increase in profit.
- Reduction in annual energy costs in various industrial sectors between 5 and 20%.
- Allocation of energy costs to individual account centers.
- For making subsequent analysis the operational and investment decision are improved.
- To increased product quality and to reduce product loss the control of operating practices are enhanced.
- For projected levels of production the calculation of energy costs are improved.
- Improved identification of maintenance requirements.

**Elements of Monitoring and Targeting System**

- **Recording:** Measuring and recording energy consumption.
- **Analyzing:** Correlating energy consumption to a measured output, such as production quantity.
- **Comparing:** Comparing energy consumption to an appropriate standard or benchmark.
- **Setting Targets:** Setting targets to reduce or control energy consumption.
- **Monitoring:** Comparing energy consumption to the set target on a regular basis.
- **Reporting:** Reporting the results including any variances from the targets which have been set.
- **Controlling:** Implementing management measures to correct any variances, which may have occurred?

Particularly M and T system will involve the following:

- Checking the accuracy of energy invoices,
- Allocating energy costs to specific departments (Energy Accounting Centres).

**Methods Monitoring and Targeting System:**

**Cumulative Sum (CUSUM):**

This procedure generates cumulative sum (CUSUM) control charts. The format of the control chart is fully customizable. The data for the subgroups can be in a single column or in multiple columns. This procedure permits the defining of stages. The target value and sigma may be estimated from the data (or a subset of the data), or a target value and sigma may be entered directly. The CUSUM chart may be used for subgroup data, or for single observations at each time point.

**Cumulative Sum (CUSUM) chart**

The CUSUM chart is used to monitor the mean of a process based on samples taken from the process at given times (hours, shifts, days, weeks, months etc.). The measurements of the samples at a given time constitute a subgroup. Rather than examining the mean of each subgroup independently, the CUSUM chart shows the accumulation of information of current and previous samples. For this reason the CUSUM chart is generally better than the X-bar chart for detecting small shifts in the mean of a process.

The CUSUM chart relies on the specification of a target value and a known or reliable estimate of the standard deviation. For this reason, the CUSUM chart is better used after process control has been established.

Control Chart Formulas Suppose we have k subgroups, each of size n. Let $x_{ij}$ represent the measurement in the $j^{th}$ sample of the $i^{th}$ subgroup. The $i^{th}$ subgroup mean is calculated using,

$$\bar{x}_i = \frac{\sum\limits_{j=1}^{n} x_{ij}}{n}$$

**Steps of CUSUM:**

1. Plot the Energy – Production graph.
2. Draw the best fit straight line.
3. Derive the equation of the line.
4. Calculate the expected energy consumption based on the equation.
5. Calculate the difference between calculated and actual energy use.
6. Compute CUSUM.
7. Plot the CUSUM graph.
8. Estimate the savings.

**For Example:** 1. Draw 20 random samples each of size 4 from Normal distribution with mean 1323 and $\sigma = 32$. Construct.

    (a) CUSUM chart (Cumulative Sum Control Chart).

**Solution:** Random Samples from Normal distribution are:

| $Xi \rightarrow N(0, 1)$ |
| --- |
| 1321.31 |
| 1305.15 |
| 1339.42 |
| 1291.5 |
| 1329.75 |
| 1330.99 |
| 1313.09 |
| 1330.37 |
| 1327.6 |
| 1317.55 |
| 1292.52 |
| 1349.66 |
| 1368.99 |
| 1309.58 |
| 1300.05 |
| 1343.17 |
| 1360 |
| 1349.39 |
| 1412.53 |
| 1341.83 |

**Example of CUSUM chart**

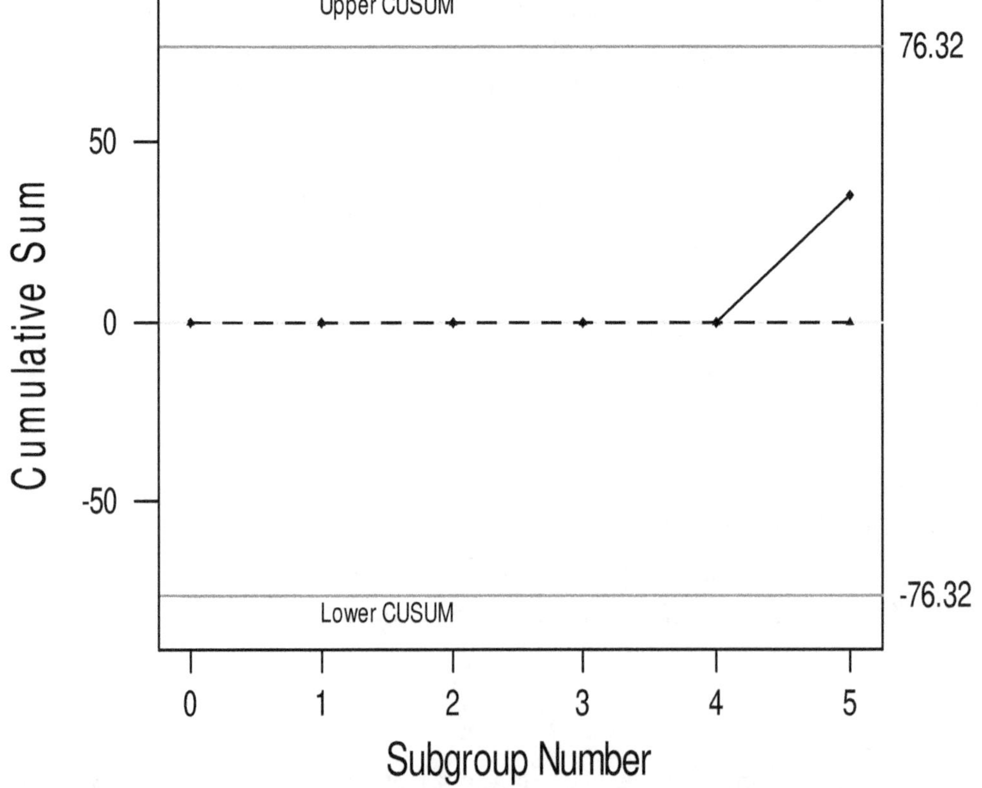


**Fig. 4.3**

**Comment:** Process in Control.

## 4.3 ENERGY MANAGEMENT AND AUDIT

### Definition and Objectives of Energy Management

**Energy management:**

The optimization and effective use of energy so as to produce goods/products at low cost and competitive prices

Or

*"The judicious and effective use of energy to maximise profits (minimise costs) and enhance competitive positions." "The strategy of adjusting and optimising energy, using systems and procedures so as to reduce energy requirements per unit of output while holding constant or reducing total costs of producing the output from these systems "*

***Source:** (Cape Hart, Turner and Kennedy, (1997) Guide to Energy Management Fairmont Press Inc).

**Objectives of energy management:**

- To reduce energy consumption.
- Effective use of energy consumption.
- Minimize energy cost.
- Reduce waste of energy.
- Reduce emission so as minimize environmental effects.
- Adopt new technology.

**Energy Audit:**

Energy audit is an energy management technique which indicates how efficiently energy is being used and indicates the energy saving areas where it is being wasted. It can also show methods to conserve energy and thus increase productivity. The energy measurement and energy monitoring are the basic steps for energy audit

Energy Audit is the systematic approach for decision making in the area of energy management. It attempts to balance the total energy inputs with its use, and serves to identify all the energy streams in a facility. It quantifies energy usage according to its discrete functions. Industrial energy audit is an effective tool in defining and pursuing comprehensive energy management programme.

The Energy Conservation Act, 2001, Energy Audit is defined as "the verification, motoring and analysis of use of energy including submission of technical report containing recommendations for improving energy efficiency with cost benefit analysis and an action plan to reduce energy consumption".

An energy Audit is the first step in energy management programme.

The objective of energy audit

- It shows how effectively or efficiently energy used.
- Highlights the areas where saving of energy cost is possible.
- Recommendations for Energy Conservation.
- Cost benefits analysis.
- It shows ways to improve productivity and efficiency of process.

**Need for Energy Audit**

1. To reduce energy consumption per unit of product output.
2. An audit program will help to keep focus on variations which occurs in the energy cost, availability and reliability of supply of energy.
3. Energy audit is the translation of conservation ideas into realities.

4.  Energy audit helps to understand more about the ways energy and fuel are used in any industry, and help in identifying the areas where waste can occur and where scope for improvement exists.

## Type of Energy Audit

The type of Energy Audit to be performed depends on:

1.  Function and type of industry.

2.  Depth to which final audit is needed.

3.  Potential and magnitude of cost reduction desired.

Thus Energy Audit can be classified into the following two types:

(i)  Preliminary Audit

(ii)  Detailed Audit

## Preliminary Energy Audit Methodology

Preliminary energy audit is a relatively quick exercise to:

- Establish energy consumption in the organization.
- Estimate the scope for saving.
- Identify the most likely (and the easiest areas for attention.
- Identify immediate (especially no-/low-cost) improvements/savings.
- Set a 'reference point'.
- Identify areas for more detailed study/measurement.
- Preliminary energy audit uses existing, or easily obtained data.

## Preliminary Audit:

It is also called as walk-through audit. It is planned 2-3 days audit.

In this audit, audit team will study the energy consumption in the organization and point out the areas where immediate saving in energy is possible like changing the faulty measurement meters, changing the location of devices etc.

Find out the opportunity of any energy conservation and suggest or recommendation for implement energy conservation like replacing all CFL bulbs by LED bulbs.

It also identifies the areas where detailed study of energy is required.

For example, from boiler exhaust gases are highly polluted. So Preliminary audit team suggested for detailed audit of boiler.

Identify immediate improvements/savings.

Identify areas for more detailed study/measurement.

Audit done on basis of data given by management.

The fundamental goal of energy management is to produce goods and provide services with the least cost and least environmental effect. The term energy management means many things to many people.

## Detailed Energy Audit Methodology

A comprehensive audit provides a detailed energy project implementation plan for a facility, since it evaluates all major energy using systems.

This type of audit offers the most accurate estimate of energy savings and cost. It considers the interactive effects of all projects, accounts for the energy use of all major equipment, and includes detailed energy cost saving calculations and project cost.

In a comprehensive audit, one of the key elements is the energy balance. This is based on an inventory of energy using systems, assumptions of current operating conditions and calculations of energy use. This estimated use is then compared to utility bill charges.

Detailed energy auditing is carried out in three phases: Phase I, II and III.

Phase I - Pre Audit Phase

Phase II - Audit Phase

Phase III - Post Audit Phase

## A Guide for Conducting Energy Audit at a Glance

Industry-to-industry, the methodology of Energy Audits needs to be flexible.

A comprehensive ten-step methodology for conduct of Energy Audit at field level is presented below. Energy Manager and Energy Auditor may follow these steps to start with and add/change as per their needs and industry types.

### Table 4.2: Ten Steps Methodology for Detailed Energy Audit

| Phase | Steps | Plan of action | Purpose/ Results |
|---|---|---|---|
| Phase I | Step-I | 1. Plan and Organise<br>2. Walk through Audit<br>3. Informal Manage with, Production / Plant Manager | Resource, Planning, Establish/organize . |
| | | | Energy Audit team. |
| | | | Organise Instruments and time frame. |
| | | | Macro data collection. |
| | | | Familiarization of process/plant activity. |
| | | | First hand observation and assessment of current level operation and practices. |
| | Step-II | Conduct of brief meeting/awareness programme with all divisional heads and concerned persons (2-3 hrs.) | Building-up cooperation. |
| | | | Issue Questionnaire for each department. |
| | | | Orientation, Awareness creation. |
| Phase II | Step-III | Primary data gathering , flow process flow diagram and energy utility diagram | Historic data analysis baseline data collection. |
| | | | Prepare process flow charts. |
| | | | All service utilizes system diagram example single line power distribution diagram water compressed air and steam distribution. |
| | | | Design operating data and schedule of operation. |
| | | | Annual energy bill and energy consumption pattern refer manual log sheet name plate interview. |
| | Step-IV | Conduct survey and monitoring | Measurements: Motor survey insulation and lightning survey with portable instruments for collection of more and accurate data confirm and compare operating data with design data. |
| | Step-V | Conduct Detail experimental trail of selected energy guzzler | Trial and experiments:<br>1. 24 hours power monitoring.<br>2. Load variance trends in palms point compressor etc.<br>3. Boiler efficiency trials for 4 to 8 hours.<br>4. Foreign and sufficiency trials equipments performance experiment, ETC. |

| | Step-VII | Analysis of energy use | Energy and material balance and energy loss/waste analysis. |
|---|---|---|---|
| | Step-VII | Identification and development of energy conservation opportunities | Identification and consolidation measures. |
| | | | Conceive develop and refined ideas. |
| | | | Review the previous ideas suggested by unit personal. |
| | | | Use brainstorming and value analysis techniques. |
| | | | Review the previous ideas suggested by energy audit. |
| | | | Contact vendors for new/ efficient technology. |
| | Step-VIII | Cost benefit analysis | Access technical feasibility economic viability and prioritization of ENCON options for implementation. |
| | | | Select the most promising projects. |
| | | | Prioritise by low medium long term measures. |
| | Step-IX | Reporting and Presentation to the Top Management | Documentation, Report Presentation to the top Management. |
| Phase-III | Step-X | Implementation and Follow- up | Assist. and Implement ENCON recommendation measures and Monitor the performance. |
| | | | Action plan, Schedule for implementation. |
| | | | Follow-up and periodic review. |

### Phase I – Pre-audit phase activities

A structured methodology to carry out an energy audit is necessary for efficient working. An initial study of the site should always be carried out, as the planning of the procedures necessary for an audit is most important.

### Initial site visit and preparation required for detailed auditing

An initial site visit may take one day and gives the Energy Auditor/Engineer an opportunity to meet the personnel concerned, to familiarize him with the site and to assess the procedures necessary to carry out the energy audit.

During the initial site visit the Energy Auditor/Engineer should carry out the following actions:

Discuss with the site's senior management the aims of the energy audit.

- Discuss economic guidelines associated with the recommendations of the audit.
- Analyse the major energy consumption data with the relevant personnel.
- Obtain site drawings where available - building layout, steam distribution, compressed air distribution, electricity distribution etc.
- Tour the site accompanied by engineering/production.

### The main aims of this visit are:

- To finalise Energy Audit team
- To identify the main energy consuming areas/plant items to be surveyed during the audit.
- To identify any existing instrumentation/ additional metering required.
- To decide whether any meters will have to be installed prior to the audit example kWh, steam, oil or gas meters.
- To identify the instrumentation required for carrying out the audit.
- To plan with time frame.

- To collect macro data on plant energy resources, major energy consuming centres.

- To create awareness through meetings/programme.

## Phase II - Detailed energy audit activities:

Depending on the nature and complexity of the site, a comprehensive audit can take from several weeks to several months to complete. Detailed studies to establish, and investigate, energy and material balances for specific plant departments or items of process equipment are carried out. Whenever possible, checks of plant operations are carried out over extended periods of time, at nights and at weekends as well as during normal daytime working hours, to ensure that nothing is overlooked.

The audit report will include a description of energy inputs and product outputs by major department or by major processing function, and will evaluate the efficiency of each step of the manufacturing process. Means of improving these efficiencies will be listed, and at least a preliminary assessment of the cost of the improvements will be made to indicate the expected payback on any capital investment needed. The audit report should conclude with specific recommendations for detailed engineering studies and feasibility analyses, which must then be performed to justify the implementation of those conservation measures that require investments.

## The information to be collected during the detailed audit includes:

- Energy consumption by type of energy, by department, by major items of process equipment, by end-use.

- Material balance data (raw materials, intermediate and final products, recycled materials, use of scrap or waste products, production of by-products for re-use in other industries etc.).

- Energy cost and tariff data.

- Process and material flow diagrams.

- Generation and distribution of site services (example. compressed air, steam).

- Sources of energy supply (for example, electricity from the grid or self-generation).

- Potential for fuel substitution, process modifications, and the use of co-generation systems (combined heat and power generation).

- Energy Management procedures and energy awareness training programs within the establishment.

**Existing baseline information and reports are useful to get consumption pattern, production cost and productivity levels in terms of product per raw material inputs. The audit team should collect the following baseline data:**

- Technology, processes used and equipment details.

- Capacity utilization.

- Amount and type of input materials used.

- Water consumption.

- Fuel Consumption.

- Electrical energy consumption.

- Steam consumption.

- Other inputs such as compressed air, cooling water etc.

- Quantity and type of wastes generated.

- Percentage rejection/reprocessing.

- Efficiencies/yield.

## Phase-III - Implementation and Follow-up

Assist and Implement ENCON recommendation measures and Monitor the performance and make action plan, Schedule for implementation Follow-up.

### Multiple Choice Questions

1. Energy monitoring and targeting is built on the principle of " ......".

   (a) "Production can be reduced to achieve reduced energy consumption".

   (b) "Consumption of energy is proportional to production rate".

   (c) "You cannot manage what you do not measure".

   (d) None of the above.

2. One of the following is not the element of energy monitoring and targeting system ......

   (a) Recording the energy consumption.

   (b) Comparing the energy consumption.

   (c) Controlling the energy consumption.

   (d) Reducing the production.

3. M and T involves a systematic, disciplines division of the facility in to energy cost centres.

   (a) True                    (b) False

4. The empirical relationship used to plot production Vs Energy consumption is (Y = energy consumed for the period; C = fixed energy consumption; M = energy consumption directly related to production; X = production for same period).

   (a) $X = Y + MC$        (b) $Y = Mx + C$        (c) $M = Cx + Y$        (d) None of above

5. 1 kg LPG = ____ kcal.

   (a) 12,000 kcal        (b) 8000 kCal        (c) 6000 kCal        (d) 4000 kCal

6. What is CUSUM?

   (a) Cumbersome                        (b) Cumulative Sum

   (c) Calculated Sum                    (d) None of these

7. A CUSUM graph follows a random fluctuation trend and oscillates around.

   (a) 100            (b) 100%            (c) 0            (d) None of the above

8. To draw a CUSUM chart following data is required ......

   (a) Monthly energy consumption and monthly production.

   (b) Monthly specific energy consumption and turn over.

   (c) Monthly profits and production.

   (d) None of these.

9. For any company, energy consumption mostly relates to ........

   (a) Profits            (b) Inventory            (c) Production            (d) All the these

10. Bureau of Energy Efficiency (BEE) establish in the year of ......

    (a) 2010	(b) 2000	(c) 2002	(d) 2001

11. Logo related with ......



**Fig. 4.4**

    (a) Bureau of Energy Efficiency (BEE)	(b) Bureau of Indian Standards

    (c) Both (a) and (b)	(d) None of the above

12. Bureau of Energy Efficiency (BEE) under the Energy Conservation Act ......

    (a) 2010	(b) 2000	(c) 2002	(d) 2001

13. ...... is the key to a systematic approach for decision-making in the area of energy management.

    (a) Audit Phase	(b) Energy Audit

    (c) Pre audit phase	(d) Conduct survey and monitoring

14. Which is the key to a systematic approach for decision-making in the area of energy management?

    (a) Audit Phase	(b) Energy Audit

    (c) Pre-audit phase	(d) Conduct survey and monitoring

15. Which of the following is false?

    (a) A structured methodology to carry out an energy audit is necessary for efficient working.

    (b) An initial site visit may take one day and gives the Energy Auditor/Engineer an opportunity to meet the personnel concerned, to familiarize him with the site and to assess the procedures necessary to carry out the energy audit.

    (c) The audit report will include a description of energy inputs and product outputs by major department or by major processing function, and will evaluate the efficiency of each step of the manufacturing process.

    (d) Process analysis is not a useful tool for process integration measures.

16. "The judicious and effective use of energy to maximise profits and enhance competitive positions". This can be the definition of ......

    (a) Energy conservation	(b) Energy management

    (c) Energy policy	(d) Energy Audit

17. The objective of energy management includes ......

    (a) Minimising energy costs	(b) Minimising waste

    (c) Minimising environmental degradation	(d) All of these

18. An energy policy does not include ......

    (a) Target energy consumption reduction

    (b) Time period for reduction

    (c) Declaration of top management commitment

    (d) Future production projection

19. The various types of the instruments, which requires during audit need to be ......

    (a)  easy to carry                          (b)  easy to operate

    (c)  inexpensive                            (d)  all of these

20. In Labels Arrow mark represent ......

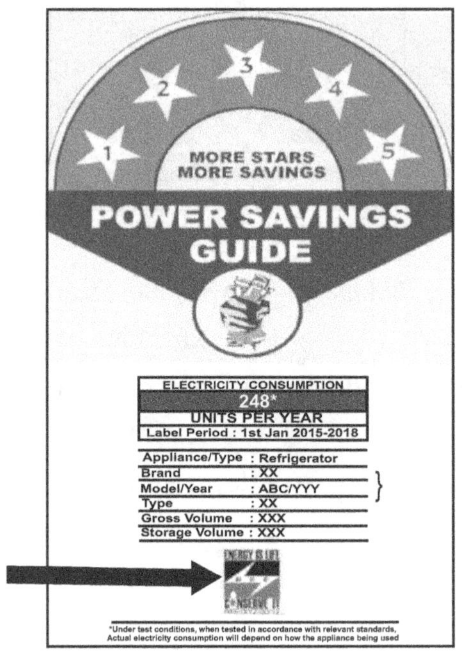


**Fig. 4.5**

    (a)  Technical Detail        (b)  Brand Detail        (c)  Efficiency parameter (d)  Unique Code

## Answers

| Question No. | 1 | 2 | 3 | 4 | 5 | 6 | 7 | 8 | 9 | 10 |
|---|---|---|---|---|---|---|---|---|---|---|
| Answer | (c) | (d) | (a) | (b) | (a) | (b) | (c) | (a) | (c) | (c) |
| Question No. | 11 | 12 | 13 | 14 | 15 | 16 | 17 | 18 | 19 | 20 |
| Answer | (b) | (d) | (c) | (a) | (a) | (b) | (d) | (d) | (d) | (d) |

# Chapter 5

# AGRICULTURE EQUIPMENT AND POST-HARVEST TECHNOLOGY

### *Weightage of Marks = 10, Teaching Hours = 06*

### *Syllabus*

5.1 Tillers, Sowing and planting equipment, Welding machines, Spraying machines, Harvesting, Post harvesting machineries.

5.2 Elements of cold chain.

5.3 National Cooling Action Plan (NCAP)

### *About this Chapter*

At the end of this chapter, students will be able to:

- Explain working of differential agricultural equipment.
- Name different elements of cold chain.
- List the features of NCAP.

## 5.1 INTRODUCTION

Plants are the primary production units of agriculture. They receive carbon dioxide from the air through their leaves, and receive water and nutrients from the soil through their roots. Using carbon dioxide, water, nutrients, and solar energy, plants produce seeds, fruits, roots, fibers, and oils that people can use. The growth of plants happens in nature without any human intervention. However, agriculture arises when people exert control over plant growth. Machines are used as an extension of people's ability to produce and care for plants.

A *crop* is a group of similar plants which are growing within the same land area.

For example, if a farm produces rice and wheat, that farm is said to produce two crops. A farmer must complete certain operations in order to successfully produce a crop. The first operation is a mechanical stirring of the soil, called *tillage*, to prepare the seed bed. The second operation is called *planting* and it places the seeds in the tilled soil at the correct depth with the appropriate spacing between seeds. When the required soil temperature and soil water content are present, the seeds will germinate and then grow leaves and roots. For some crops the seeds are planted in a small area called a nursery and then the small plants are transplanted to the fields where they will grow to maturity.

As the plants grow the farmer must protect them from pests such as weeds (unwanted plants), insects, other animals, and diseases. Mechanical cultivation (tillage between the plants) is used to control weeds in some cases. Chemicals are frequently used to control weeds, insects, and diseases. Fences and/or noise-making devices may be used for protection from larger animals. The final crop production operation is the *harvesting* of the plant parts which have economic value for the farmer. In some cases, more than one part of the plant may have economic value. For example, a farmer may use rice straw (stems and leaves) as an energy resource after the rice seeds have been removed from the plants. In other cases, the crop residue (unused plant parts) is stirred into the soil during tillage for the next crop.

## 5.2 FARM MECHANIZATION

**Mechanized agriculture** is the process of using agricultural machinery to mechanize the work of agriculture, greatly increasing farm worker productivity. The effective mechanization contributes to increase production in two major ways: firstly the time lines sof operation and secondly the good quality of work. The requirement of power for certain operations like seedbed preparation, cultivation and harvesting becomes so great that the existing human and animal power in the country appears to be inadequate. As a result, the operations are either partially done or sometimes completely neglected, resulting in low yield due to poor growth or untimely harvesting or both.

## 5.3 SCOPE OF MECHANIZATION

It is quite true that the Indian farmers have the lowest earnings per capita because of the low yield per hectare they get from their holdings. One of the few important means of increasing farm production per hectare is to mechanize it. Mechanization in India may have to be done at various levels. Broadly, it can be done in three different ways:

(i) By introducing the improved agricultural implements on small size holdings to be operated by bullocks.

(ii) By using the small tractors, tractor-drawn machines and power tillers on medium holdings to supplement existing sources.

(iii) By using the large size tractors and machines on the remaining holdings to supplement animal power source.

History indicates that the development in farm mechanization is very closely related to the shortage of human labour and industrial development in the country. Farmers of India like their counter parts in other countries are interested to improve their income, lifestyle and general well being. The agriculture and allied sector continues to be significant for the inclusive and sustainable growth of the Indian economy. Indian Agriculture Sector not only ensures food security but also provides employment for substantial volume of population, directly and indirectly. Though agriculture contributed only 17.4% to the country's Gross Value Added for the year 2016-17 (at current prices), still it is a driver for demand creation.

According to the World Bank estimates, half of the Indian population would be urban by the year 2050. It is estimated that percentage of agricultural workers in total work force would drop to 25.7% by 2050 from 58.2% in 2001. Thus, there is a need to enhance the level of farm mechanization in the country. Due to intensive involvement of labour in different farm operations, the cost of production of many crops is quite high. Human power availability in agriculture also increased from about 0.043KW/ ha in 1960-61 to about 0.077 KW/ ha in 2014- 15. However, as compared to tractor growth, increase in human power in agriculture is quite slow. Over the year, the shift has been towards the use of mechanical and electrical sources of power. In 1960-61, about 93% farm power was coming from animate sources, which has reduced to about 10% in 2014-15. On the other hand, mechanical and electrical sources of power have increased from 7% to about 90% during the same period. As there is predominance of small operational holding in Indian Agriculture, it is, therefore, necessary to consolidate the land holdings to reap the benefits of agricultural mechanization.

1.  **Preparing land for planting:**

**Fig. 5.1: Preparing Land**

## 2. Seed drilling, planting:

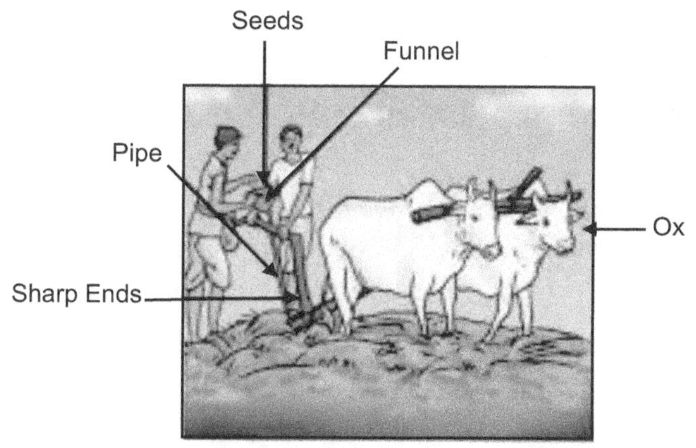


**Fig. 5.2: Traditional Method of Sowing**

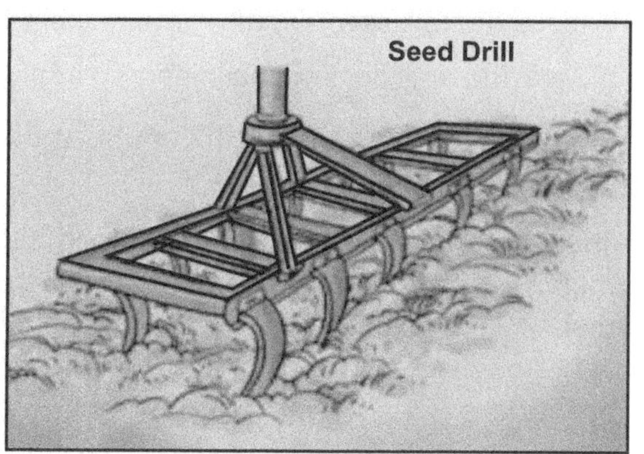


**Fig. 5.3: Seed Drill**

## 3. Weeding, crop spraying:

**Fig. 5.4: Crop Sprayin**

**Fig. 5.5: Weeding**

## 4. Harvesting:

**Fig. 5.6: Harvesting**

**Fig. 5.7: Conventional Harvesting**

The three ways in which progress can be made towards increased power, all of which must be worked on simultaneously in combination with integrated and matched implements.

1. By improving bullock harness and hitches;

2. By developing and introducing small tractors;

3. By increasing the number of large horse power tractors.

These machines will be helpful in providing power efficiently for good seedbed preparation, which is quite essential for maximizing the germination of the seed and seedling growth. In addition to this, the mechanization of the following fields of agriculture-is equally essential.

(i) Shaping and levelling of farm fields for getting even distribution and depth of irrigation water;

(ii) Development of planting and fertilizing machines to place these edm or eprecisely in rows and to place the fertilizer correctly with respect to seed or plant;

(iii) Spraying and dusting machinery to apply chemicals for weed and pest control;

After production operations such as harvesting, threshing, winnowing and drying.

## Agricultural Equipments:

Agricultural machinery is machinery used in farming or other agriculture. There are many types of such equipment, from hand tools and power tools to tractors and the countless kinds of farm implements that they tow or operate. Diverse arrays of equipment are used in both organic and non-organic farming. Especially since the advent of mechanised agriculture, agricultural machinery is an indispensable part of how the world is fed.

Agricultural equipment is any kind of machinery used on a farm to help with farming. The best-known example of this kind is the tractor.

## Tractor:

- Tractor is a word which is derived from TRACTION and MOTOR.

- Tractor is a vehicle specifically designed to deliver a high tractive effort at slow speeds, for the purposes of hauling a trailer or machinery used in agriculture or construction.

- It should withstand load when tractor is working in field with implements in uneven fields.

- It should have a provision for running other stationary machines like Threshers, Fertilizer, and Sprayers etc. that is it should have PTO shaft arrangement.

- It should have suitable hydraulic system for handling implements or equipments.

## Classification of Tractors:

## On the basis of structural-design:

(i) **Wheel tractor:** Tractors, having three of four pneumatic wheels are called wheel tractors. Four-wheel tractors are most popular everywhere.

**Fig. 5.8: Wheel Tractor**

(ii) **Crawler tractor:** This is also called track type tractor or chain type tractor. In such tractors, there is endless chain or track in place of pneumatic wheels.

**Fig. 5.9: Crawler Tractor**

**(iii) Walking tractor (Power tiller):** Power tiller is a walking type tractor. This tractor is usually fitted with two wheels only. The direction of travel and its controls for field operation is performed by the operator, walking behind the tractor.

**Fig. 5.10: Power Tiller**

**On the basis of purpose:**

**(a) General purpose tractor:** It is used for major farm operations; such as ploughing, harrowing, sowing, harvesting and transporting work. Such tractors have (i) low ground clearance (ii) increased engine power (iii) good adhesion and (iv) wide tyres.

**Fig. 5.11: General Purpose Tractor**

**(b) Row crop tractor:** It is used for crop cultivation. Such tractor is provided with replaceable driving wheels of different tread widths. It has high ground clearance to save damage of crops. Wide wheel track can be adjusted to suit inter row distance.

**Fig. 5.12: Row Crop Tractor**

**(c) Special purpose tractor:** It is used for definite jobs like cotton fields, marshy land, hillsides, garden etc. Special designs are there for special purpose tractor.

**Fig. 5.13: Special Purpose Tractor**

## Tractor Components:

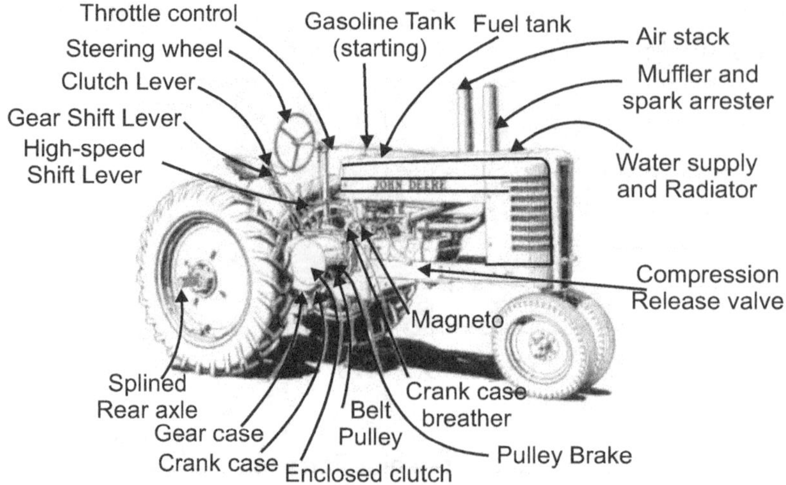



**Fig. 5.14: Tractor Components**

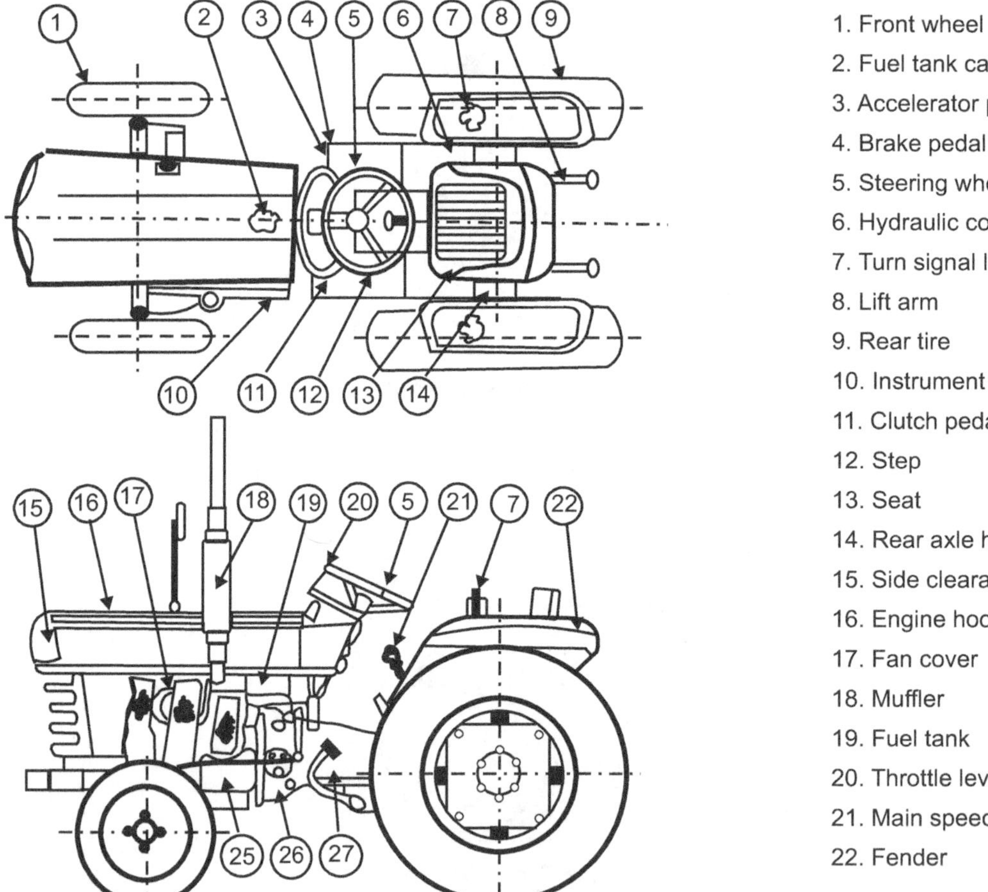



1. Front wheel
2. Fuel tank cap
3. Accelerator pedal
4. Brake pedal
5. Steering wheel
6. Hydraulic control lever
7. Turn signal lamp
8. Lift arm
9. Rear tire
10. Instrument panel
11. Clutch pedal
12. Step
13. Seat
14. Rear axle housing
15. Side clearance
16. Engine hood
17. Fan cover
18. Muffler
19. Fuel tank
20. Throttle lever
21. Main speed change lever
22. Fender

**Fig. 5.15: Front View and Side View of Tractor**

## Control Board or Dash Board of a Tractor:

The control board of a tractor generally consists of:

(1) Main switch

(2) Throttle lever

(3) Decompression lever

(4) Hour meter

(5) Light switch

(6) Horn button

(7) Battery charging indicator

(8) Oil pressure indicator and

(9) Water temperature gauge.

**Tractor Tyres and Front Axle:**

**Tyres:**

The tyres are available in many sizes with the ply ratings as 4, 6 or 8. The ply rating of tyres indicates the comparative strength of tyres. The higher the rating, the stronger are the tyres. The tyres size 12—38 means, that the sectional diameter of tyres is 12" and it is mounted on a rim of 38" diameter. The inflation pressure in the rear wheels of the tractor varies between 0.8 to 1.5 kg/cm$^2$. The inflation pressure of the front wheel varies from 1.5 to 2.5 kg/cm$^2$. Useful life of the pneumatic tyres under normal operating condition may be about 6000 working hours for draw bar work.

**Fig. 5.16: Tractor Tyres**

**Front Axle:** Front axle is the unit on which front wheel is mounted. This wheel is an idler wheel by which tractor is steered in various directions. The axle is a rigid tubular or I-section steel construction pivoted at the centre. There are various adjustments of front wheel.

**Hitching System of Tractor Drawn Implements:**

Tractor drawn implements possess higher working capacity and are operated at higher speeds. These implements need more technical knowledge for operations and maintenance work. Tractor drawn implements may be (a) Trailed type (b) Semi-mounted type and (c) Mounted type.

(a) **Trailed type implement:** It is one that is pulled and guided from single hitch point but its weight is not supported by the tractor.

(b) **Semi-mounted type implement:** This type of implement is one which is attached to the tractor along a hinge axis and not at a single hitch point. It is controlled directly by tractor steering unit but its weight is partly supported by the tractor.

(c) **Mounted type implement:** A mounted implement is one which is attached to the tractor, such that it can be controlled directly by the tractor steering unit. The implement is carried fully by the tractor when out of work.

**Some Important Terms Connected With Tractors:**

**Wheel base:** Wheel base is the horizontal distance between the front and rear wheels of a tractor, measured at the ground contact.

**Ground clearance:** It is the height of the lowest point of die tractor from the ground surface, the tractor being loaded to its maximum permissible weight.

**Track:** Track is the distance between the two wheels of the tractor on the same axle, measured at the point of ground contact.

**Turning space:** It is the diameter of the smallest circle, described by the outermost point of the tractor, while moving at a speed, not exceeding 2 km/hr with the steering wheels in full lock.

(a) **Cage wheel:** It is a wheel or an attachment to a wheel with spaced cross bars for improving the traction of the tractor in a wet field. It is generally used in paddy fields.

## Selection of Tractor:

1. **Land holding:** Under a single cropping pattern, it is normally recommended to consider 1 hp for every 1 hectares of land. In other words, one tractor of 20-25 hp is suitable for 20 hectares farm.

2. **Cropping pattern:** Generally less than 1.0 hectare/hp have been recommended where adequate irrigation facilities are available and more than one crop is taken. So a 30-35 hp tractor is suitable for 25 hectares farm.

3. **Soil condition:** A tractor with less wheel base, higher ground clearance and low overall weight may work successfully in lighter soil but it will not be able to give sufficient depth in black cotton soil.

4. **Climatic condition:** For very hot zone and desert area, air cooled engines are preferred over water-cooled engines. Similarly for higher altitude, air cooled engines are preferred because water is liable to be frozen at higher altitude.

5. **Repairing facilities:** It should be ensured that the tractor to be purchased has a dealer at nearby place with all the technical skills for repair and maintenance of machine.

6. **Running cost:** Tractors with less specific fuel consumption should be preferred over others so that running cost may be less.

7. **Initial cost and resale value:** While keeping the resale value in mind, the initial cost should not be very high; otherwise higher amount of interest will have to be paid.

8. **Test report:** Test report of tractors released from farm machinery testing stations should be consulted for guidance.

## Tillers:

A power tiller is a two-wheeled agricultural implement, fitted with rotary tillers, that is fast gaining traction in the Indian agricultural scene for its multi-purpose uses. In addition to tilling the soil, power tillers can be used for ploughing the soil, sowing seeds, planting seedlings, adding fertilizers, spraying fertilizers, herbicides, and water, pumping water, harvesting crops, threshing crops, and transporting crops. All these additional tasks are performed by attaching agricultural devices such as ploughs, seeders, planters, pumps, sprays, harvesters, threshers, and carriers to the power tiller.

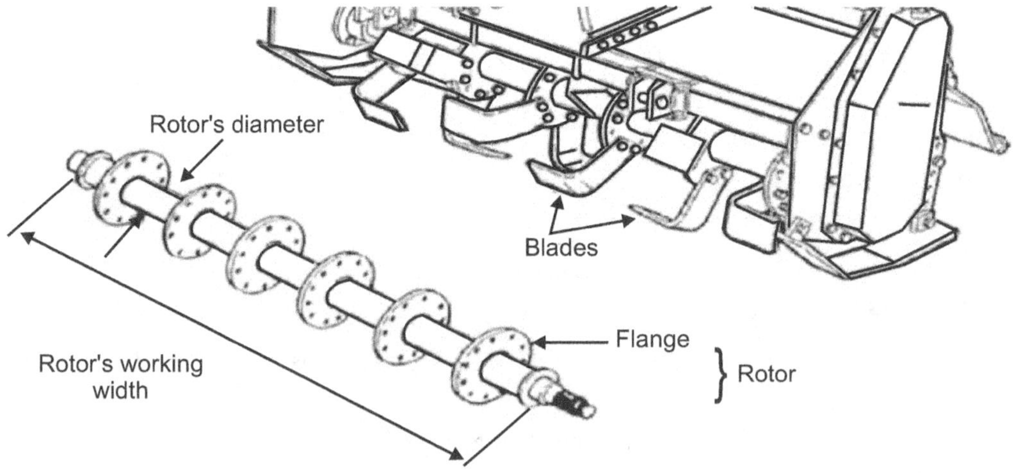


**Fig. 5.17: Tillers**

## Working of Tillers:

Power tillers can work on small as well as large farms, but they are particularly recommended for farms where the small land size or the rugged, sloping terrain might make manoeuvring a tractor about impractical. Difficult, or even dangerous. In hilly regions in India, where only terrace farming is possible, farmers have found power tillers to be extremely useful.

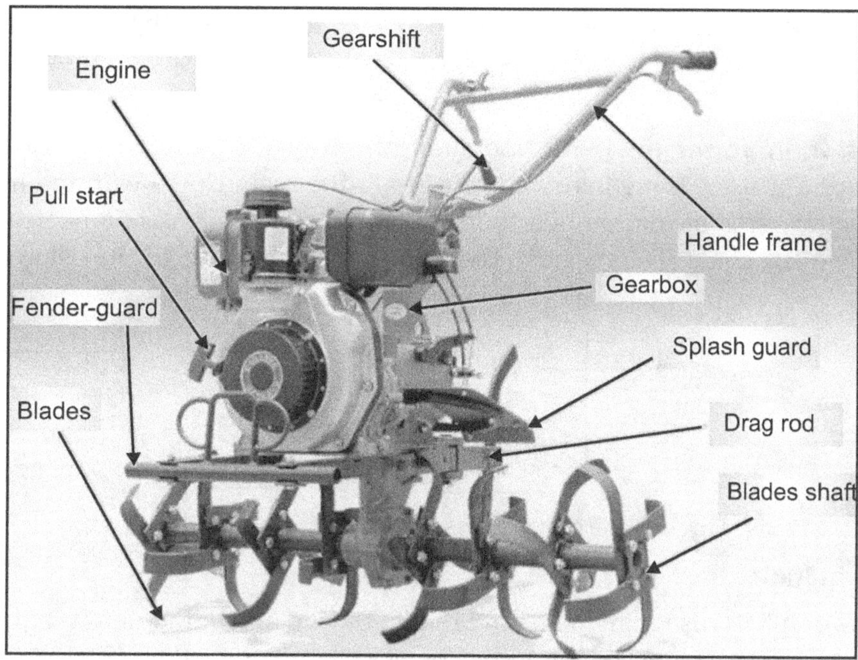



**Fig. 5.18: Components of Tillers**

Rice and paddy farmers with small farm plots use power tillers to successfully till water-flooded fields. Unlike tractors, which are heavy and can compact the soil with their weight and till deeper than necessary, power tillers till only to the required depth, softening and loosening the soil as they do, and making it easier to plant rice and paddy in the water submerged fields without any danger of drowning the plants under water.

**Types of Power Tillers:**

Power tillers are available in three different sizes :

- **Mini-sized Power Tiller:** These lightweight implements are good for vegetable and fruit farms, where the cultivation areas are small, the soil is relatively loose and does not require powerful tilling, and the crops are planted in narrow rows that need to be manoeuvred through without destroying them.

- **Medium-sized Power Tiller:** These power tillers are suitable for farms where the soil is harder and rockier. These tillers are propelled forward by the spinning tines.

- **Large-sized Power Tiller:** These power tillers are suitable for more extensive farm lands where there is sufficient space to manoeuvre the tiller as required. These tillers are propelled forward by the wheels and the spinning tines are positioned between the wheels.

**Fig. 5.19: Large-sized Power Tiller**

To select the right power tiller, farmers should consider the size of their land, the type of soil, the amount of tilling necessary, the storage space they have for the power tiller, and their budget for the power tiller.

**Power transmission in power tiller:** For operation of power tiller, the power is obtained from the IC Engine, fitted on the power tiller. The engine power goes to the main clutch with the help of belt or chain. From main clutch, the power is divided in two routes, one goes to transmission gears, steering clutch and then to the wheel. The other component goes to the tilling clutch and then to the tilling attachment.

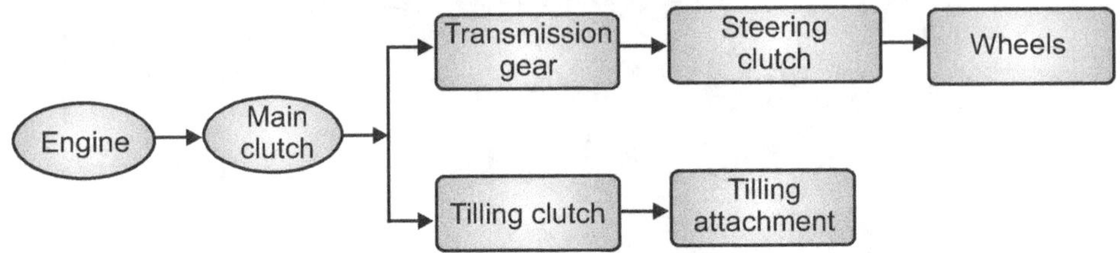


**Fig. 5.20: Power transmission in power tiller**

## Power Tiller Brands in India:

Most of the power tillers used in India are made by Indian and Chinese agricultural equipment manufacturers. Some of the more well-known Indian power tiller manufacturers include .

- KMW Kirloskar Power Tillers
- Kubota Power Tillers
- Kranti Power Tillers
- VST Shakti Power Tillers
- Bull Agro Power Tillers
- Southern Agro Engines Power Tillers
- KAMCO Limited Power Tillers

## Benefits of Power Tiller to Indian Farmers:

Indian farmers with small and medium sized land-holdings are likely to find power tillers more useful than tractors for the following reasons:

- Power tillers are less expensive than tractors, and the Indian government provides subsidies to Indian farmers to purchase them.
- Power tillers require less room for storage.
- Power tillers consume lower fuel amounts than tractors. If a tractor requires two litres of diesel for an hour's work, a power tiller can make do with one litre in the same time.
- Power tillers have relatively low maintenance costs as compared to tractors. Also, if farmers are diligent about cleaning the power tillers daily, replacing engine oil as required, and maintaining necessary fuel levels, the power tillers will remain in reasonably good condition for a longer period.
- Power tillers are easier to manoeuvre than tractors, and this is a very important consideration on small-sized farms and hilly terrain farms.
- Power tillers are useful in regions where there is acute labour shortage. They are useful too in regions where there is no labour shortage – the labour can be used for other, less cumbersome jobs. Also, power tillers will work faster and better than human and animal labour.
- Power tillers can be effectively used on different types of terrains – flat lands, hilly terraces, water submerged fields, and dry fields.
- Power tillers are ideal for small areas where the plants are planted in narrow rows, as the machines can be easily manoeuvred between these rows without damaging or destroying the plants.

**Tillage:**

- Tillage operations in various forms have been practiced from the very inception of growing plants. Primitive man used tools to disturb the soils for placing the seeds.

- It is a mechanical manipulation of soil to provide favourable condition for crop production. Soil tillage consists of breaking the compact surface of earth to a certain depth and to loosen the soil mass, so as to enable the roots of the crops to penetrate and spread into the soil.

- The word tillage is derived from 'Anglo-Saxon' words Tilian and Teolian, meaning 'to plough and prepare soil for seed to sow, to cultivate and to raise crops'. Jethrotull, who is considered as father of tillage suggested that thorough ploughing is necessary so as to make the soil into fine particles.

- Tilth is the physical condition of soil obtained out of tillage (or) it is the result of tillage. The tilth may be a coarse tilth, fine tilth or moderate tilth.

**History:**

Tilling was first performed via human labor, sometimes involving slaves. Hoofed animals could also be used to till soil by trampling, in addition to pigs, whose natural instincts are to root the ground regularly if allowed to. The wooden plow was then invented. It could be pulled with human labour, or by mule, ox, elephant, water buffalo, or similar sturdy animal. Horses are generally unsuitable, though breeds such as the Clydesdale were bred as draft animals. The steel plow allowed farming in the American Midwest, where tough prairie grasses and rocks caused trouble. Soon after 1900, the farm tractor was introduced, which eventually made modern large-scale agriculture possible.

**Objectives of tillage:**

- To prepare a good seed bed, suitable for different type of crops.
- To create conditions in the soil suited for better growth of crops.
- To destroy and prevent weeds.
- To aerate the soil for proper growth of crops.
- To make the soil capable for absorbing more rain water.
- To mix-up the manure and fertilizers uniformly in the soil.
- To destroy the insects, pests and their breeding places.
- To reduce the soil erosion.
- To remove the hard pan and to increase the soil depth.

**Types of tillage:**

Tillage operations may be grouped into

1. On season tillage.                    2. Off-season tillage.

**1. On-season tillage:**

Tillage operations that are done for raising crops in the same season or at the onset of the crop season are known as on-season tillage. They may be preparatory cultivation and after cultivation.

**(A) Preparatory tillage:**

This refers to tillage operations that are done to prepare the field for raising crops. It consists of deep opening and loosening of the soil to bring about a desirable tilth as well as to incorporate or uproot weeds and crop stubble when the soil is in a workable condition. Types of preparatory tillage

(a) Primary tillage.                    (b) Secondary tillage.

**(a) Primary tillage:** The tillage operation that is done after the harvest of crop to bring the land under cultivation is known as primary tillage or ploughing. Ploughing is the opening of compact soil with the help of different ploughs. Country plough, mould board plough, bose plough, tractor and power tiller drawn implements are used for primary tillage.

**(b) Secondary tillage:** The tillage operations that are performed on the soil after primary tillage to bring a good soil tilth are known as secondary tillage. Secondary tillage consists of lighter or finer operation which is done to clean the soil, break the clods and incorporate the manure and fertilizers. Harrowing and planking is done to serve those purposes. Planking is done to crush the hard clods, level the soil surface and to compact the soil lightly. Harrows, cultivators, Guntakas and spade are used for secondary tillage.

**(c) Layout of seed bed:** This is also one of the components of preparatory tillage. Levelling board, buck scrapers etc. are used for leveling and markers are used for layout of seedbed.

## (B) After cultivation (Inter tillage):

The tillage operations that are carried out in the standing crop after the sowing or planting and prior to the harvesting of the crop plants are called after tillage. This is also called as inter cultivation or post seeding/planting cultivation. It includes harrowing, hoeing, weeding, earthing up, drilling or side dressing of fertilizers etc. Spade, hoe, weeders etc. are used for inter cultivation.

## 2. Off-season tillage:

Tillage operations done for conditioning the soil suitably for the forthcoming main season crop are called off-season tillage. Off season tillage may be, post-harvest tillage, summer tillage, winter tillage and fallow tillage. Special purpose tillage: Tillage operations intended to serve special purposes are said to be special purpose tillage.

**(a) Sub-soiling:** To break the hard pan beneath the plough layer, special tillage operation (chiseling) is performed to reduce compaction. Sub-soiling is essential and once in four to five years where heavy machineries are used for field operations, seeding, harvesting and transporting. Advantages of sub-soiling are, greater volume of soil may be obtained for cultivation of crops, excess water may percolate downward to recharge the permanent water table, reduce runoff and soil erosion and roots of crop plants can penetrate deeper to extract moisture from the water table.

**(b) Clean tillage:** It refers to working of the soil of the entire field in such a way no living plant is left undisturbed. It is practiced to control weeds, soil borne pathogen and pests.

**(c) Blind tillage:** It refers to tillage done after seeding or planting the crop (in a sterile soil) either at the pre-emergence stage of the crop plants or while they are in the early stages of growth so that crop plants (sugarcane, potato etc.) do not get damaged, but, extra plants and broad leaved weeds are uprooted.

**(d) Dry tillage:** Dry tillage is practiced for crops that are sown or planted in dry land condition having sufficient moisture for germination of seeds. This is suitable for crops like broadcasted rice, jute, wheat, oilseed crops, pulses, potato and vegetable crops. Dry tillage is done in a soil having sufficient moisture (21-23%). The soil becomes more porous and soft due to dry tillage. Besides, the water holding capacity of the soil and aeration are increased. These conditions are more favourable for soil micro-organisms.

**(e) Wet tillage or puddling:** The tillage operation that is done in a land with standing water is called wet tillage or puddling. Puddling operation consists of ploughing repeatedly in standing water until the soil becomes soft and muddy. Puddling creates an impervious layer below the surface to reduce deep percolation losses of water and to provide soft seed bed for planting rice. Puddling is done in both the directions for the incorporation of green manures and weeds. Wet tillage destroys the soil structure and the soil particles that are separated during puddling settle later. Wet tillage is the only means of land preparation for transplanting semi-aquatic crop plant such as rice. Planking after wet tillage makes the soil level and compact. Puddling hastens transplanting operation as well as establishment of seedlings. Wet land ploughs or worn out dry land ploughs are normally used for wet tillage

### Some other types of Tillage

**Minimum Tillage:** It is the minimum soil manipulation necessary to meet tillage requirements for crop production.

**Strip Tillage:** It is a tillage system in which only isolated bands of soil are tilled.

**Rotary Tillage:** It is the tillage operations employing rotary action to cut, break and mix the soil.

**Mulch Tillage:** It is the preparations of soil in such a way that plant residues or other mulching materials are specially left on or near the surface.

**Combined Tillage:** Operations simultaneously utilizing two or more different types of tillage tools or implements to simplify, control or reduce the number of operations over a field are called combined tillage.

### Difference between Tools/Implements/Machines:

**Tool:** It is an individual working element such as disc or shovel.

**Implement:** It is equipment generally having no driven moving parts, such as harrow or having only simple mechanism such as plough.

**Machine:** It is a combination of rigid or resistant bodies having definite motions and capable of performing useful work.

### Implements for Primary Tillage:

**Plough:** Ploughing is the primary tillage operations, which are performed to cut, break and invert the soil partially or completely. Ploughing essentially means opening the upper crust of the soil, breaking the clods and making the soil suitable for sowing seeds.

### Country or Indigenous Plough:

It penetrates into the soil and breaks it open. The functional components include share, body, shoe, handle and beam. It can be used for dry land, garden land and wetland ploughing operations.

**Share:** It is the working part of the plough attached to the shoe with which it penetrates into the soil and breaks it open.

**Shoe:** It supports and stabilizes the plough at the required depth.

**Body:** It is main part of the plough to which the shoe, beam and handle are generally attached. In country plough body and shoe are integral part.

**Beam:** It is generally a long wooden piece, which connects the main body of the plough to the yoke.

**Handle:** A wooden piece vertically attached to the body to enable the operator to control the plough.

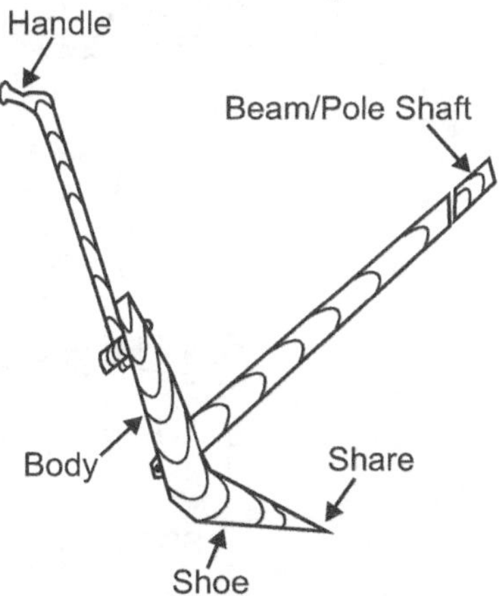


**Fig. 5.21: Plough**

**Operational Adjustments:**

- Lowering or raising the beam with respect to the plough body, resulting in a change in the angle of the share with the horizontal plane to increase or decrease the depth of operation.

- Changing the length of the beam (body to yoke on the beam) to increase or decrease the depth of operation.

- The size of the plough is represented by the width of the body.

**Mouldboard Plough:**

**Function:**

(1) Cutting the furrow slice,

(2) Lifting the soil,

(3) Turning the furrow slice, and

(4) Pulverizing the soil.

**Components:**

**Share:** It penetrates into the soil and makes a horizontal cut below the soil surface. It is a sharp, well-polished and pointed component. The shares are made of chilled cast iron or steel. The steel mainly contains about 0.70 to 0.80% carbon and about 0.50 to 0.80% manganese besides other minor elements.

**Mould board:** The mould board is that part of the plough which receives the furrow slice from the share. If lifts, turns and breaks the furrow slice. To suit different soil conditions and crop requirements, mould board has been designed in different shapes. The mould board is of following types: a) General purpose b) Stubble c) Sod or Breaker and d) Slat.

**(i) General purpose:** It is a mould board having medium curvature lying between stubble and sod. The sloping of the surface is gradual. It turns the well-defined furrow slice and pulverises the soil thoroughly. It has a fairly long mould board with a gradual twist, the surface being slightly convex.

**(ii) Stubble type:** It is short but broader mould board with a relatively abrupt curvature which lifts breaks and turns the furrow slice used in stubble soils. Its curvature is not gradual but it is abrupt along the top edge. This causes the furrow slice to be thrown off quickly, pulverising it much better than other types of mould board. This is best suited to work in stubble soil that is under cultivation for years together. Stubble soil is that soil in which stubble of the plants from the previous crop is still left on the land at the time of ploughing.

**(iii) Sod or Breaker type:** It is a long mould board with gentle curvature which lifts and inverts the unbroken furrow slice. It is used in tough soil of grasses. It turns over thickly covered soil. This is very useful where complete inversion of soil is required by the farmer.

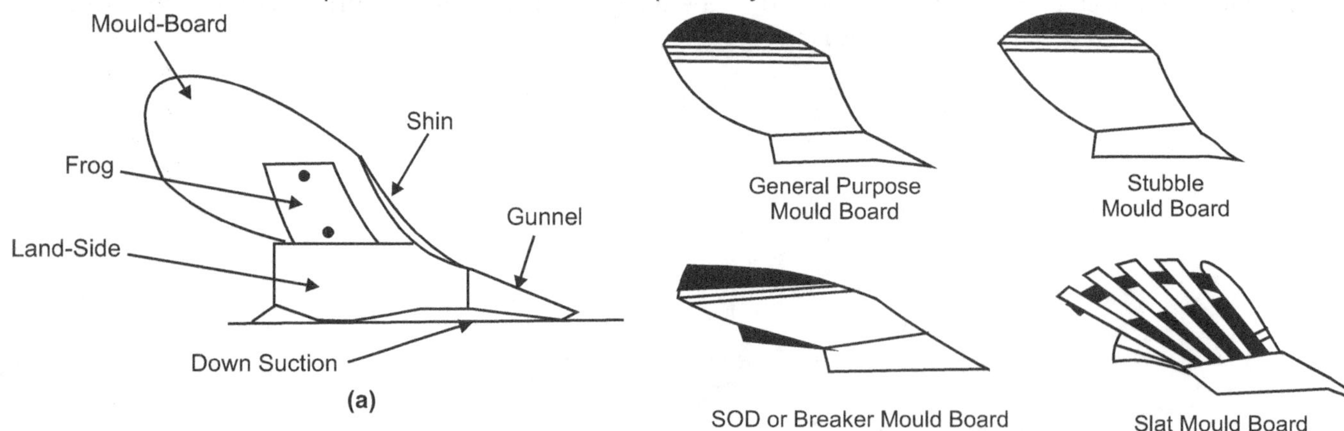

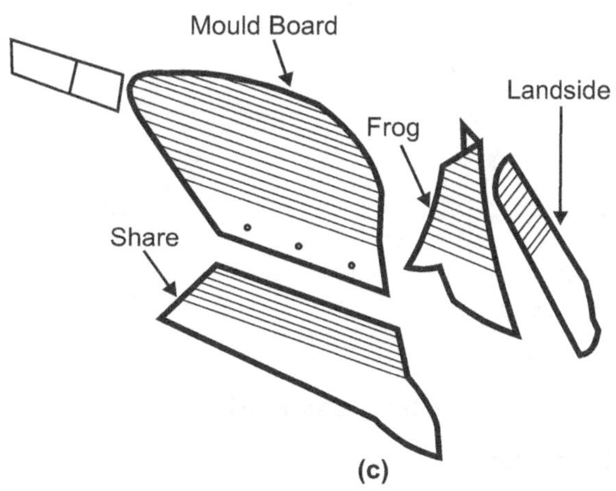


**Fig. 5.22: Type of Plough**

**(iv) Slat type:** It is a mould board whose surface is made of slats placed along the length of the mould board, so that there are gaps between the slats. This type of mould board is often used, where the soil is sticky, because the solid mould board does not scour well in sticky soils.

**(c) Land side:** It is the flat plate which bears against and transmits lateral thrust of the plough bottom to the furrow wall. It helps to resist the side pressure exerted by the furrow slice on the mould board. It also helps in stabilizing the plough while it is in operations.

**(d) Frog:** Frog is that part of the plough bottom to which the other components of the plough bottom are attached. It is an irregular piece of metal. It may be made of cast iron for cast iron ploughs or it may be welded steel for steel ploughs.

**(e) Tail piece:** It is an important extension of mould board which helps in turning a furrow slice.

**Plough Accessories:**

There are a few accessories necessary for plough such as (i) Jointer (ii) Coulter (iii) Gauge wheel (iv) Land wheel and (v) Furrow wheel.

**Jointer:** It is a small irregular piece of metal having a shape similar to an ordinary plough bottom. It looks like a miniature plough. Its purpose is to turn over a small ribbon like furrow slice directly in front of the main plough bottom. This small furrow slice is cut from the left and upper side of the main furrow slice and is inverted so that all trashes on the top of the soil are completely turned down and buried under the right hand corner of the furrow.

**Coulter:** It is a device used to cut the furrow slice vertically from the land ahead of the plough bottom. It cuts the furrow slice from the land and leaves a clear wall. It also cut strashes which are covered under the soil by the plough. The coulter may be (a) Rolling type disc coulter or (b) Sliding type knife coulter.

**Gauge wheel:** It is an auxiliary wheel of an implement to maintain uniform depth of working. Gauge wheel helps to maintain uniformity in respect of depth of ploughing in different soil conditions. It is usually placed in hanging position.

**Land wheel:** It is the wheel of the plough, which runs on the ploughed land.

**Front furrow wheel:** It is the front wheel of the plough, which runs in the furrow.

**Rear furrow wheel:** It is the rear wheel of the plough, which runs in the furrow.

**Adjustment of Mould Board Plough:**

**Vertical Suction (Vertical clearance):**

It is the maximum clearance under the land side and the horizontal surface when the plough is resting on a horizontal surface in the working position. It is the vertical distance from the ground, measured at the joining point of share and land side. It helps the plough to penetrate into the soil to a proper depth. This clearance varies according to the size of the plough.

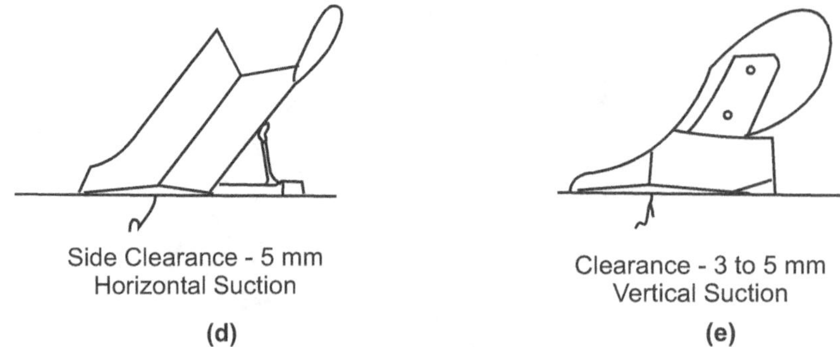



**Fig. 5.22: Vertical and Horizontal Suction**

### Horizontal Suction (Horizontal Clearance):

It is the maximum clearance between the land side and a horizontal plant touching point of share at its tunnel side and heal of land side (Fig. 5.22). This suction helps the plough to cut the proper width of furrow slice. This clearance varies according to the size of the plough. It is also known as side clearance.

### Throat clearance:

It is the perpendicular distance between point of share and lower position of the beam of the plough.

### Plough Size:

**Disc Plough:** It is a plough, which cuts, turns and in some cases breaks furrow slices by means of separately mounted large steel discs. A disc plough is designed with a view of reduce friction by making a rolling plough bottom instead of sliding plough bottom. A disc plough works well in the conditions where mould board plough does not work satisfactorily.

### Advantages of disc plough:

- A disc plough can be forced to penetrate into the soil which is too hard and dry for working with a mould board plough.
- It works well in sticky soil in which a mould board plough does not scour.
- It is more useful for deep ploughing.
- It can be used safely in stony and stumpy soil without much danger of breakage.
- A disc plough works well even after a considerable part of the disc is worn-off in abrasive soil.
- It works in loose soil also (such as peat) without much clogging.

### Disadvantages of disc plough:

- It is not suitable for covering surface trash and weeds as effectively as mould board plough does.
- Comparatively, the disc plough leaves the soil in rough and more cloddy condition than that of mould board plough.
- Disc plough is much heavier than mould board plough for equal capacities because penetration of this plough is affected largely by its weight rather than suction. There is one significant difference between mould board plough and disc plough i.e. mould board plough is forced into the ground by the suction of the plough, while the disc plough is forced into the ground by its own weight.

### Types of Disc Plough:

Disc ploughs are of two types (i) Standard disc plough and (ii) Vertical disc plough.

#### Standard disc plough:

It consists of steel disc of 60 to 90 cm diameter, set at a certain angle to the direction of travel. Each disc revolves on a stub axle in a thrust bearing, carried at the lower end of a strong stand which is bolted to the plough beam. The angle of the disc to the vertical and to the furrow wall is adjustable. In action, the disc cuts the soil, breaks it and pushes it sideways. There is little inversion of furrow slice as well as little burying of weeds and trashes. The disc plough may be mounted type or trailed type. In mounted disc plough, the side thrust is taken by the wheels of the tractor. Disc is made of heat treated steel of 5 mm to 10 mm thickness. The amount of concavity varies with the diameter of the disc. The approximate values being 8 cm for 60 cm diameter disc and 16 cm for 95 cm diameter. A few important terms connected with disc plough is explained below.

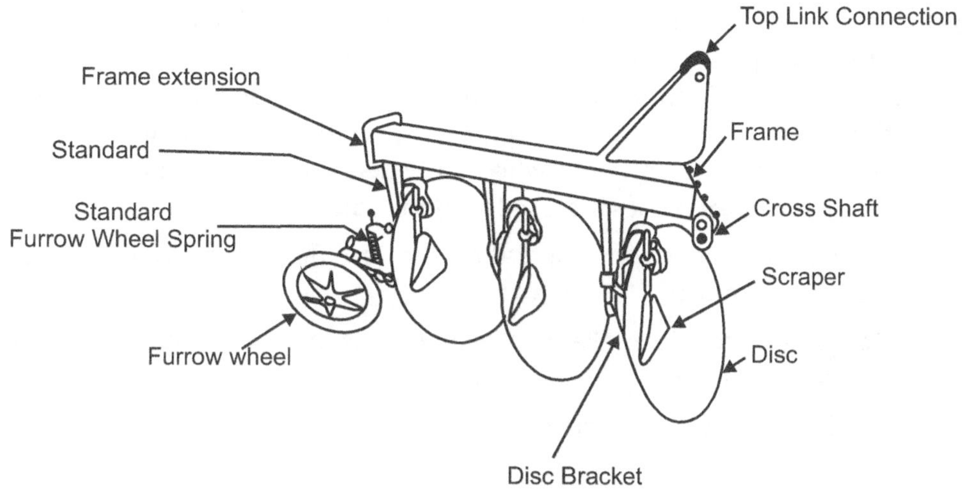


**Fig. 5.23: Standard Disc Plough**

**Disc:** It is a circular, concave revolving steel plate used for cutting and inverting the soil.

**Disc angle:** It is the angle at which the plane of the cutting edge of the disc is inclined to the direction of travel. Usually the disc angle of good plough varies between 42° to 45°.

**Tilt angle:** It is the angle at which the plane of the cutting edge of the disc is inclined to a vertical line. The tilt angle varies from 15° to 25° for a good plough.

**Scraper:** It is a device to remove soil that tends to stick to the working surface of a disc.

The size of the mouldboard plough is expressed by width of cut of the soil.

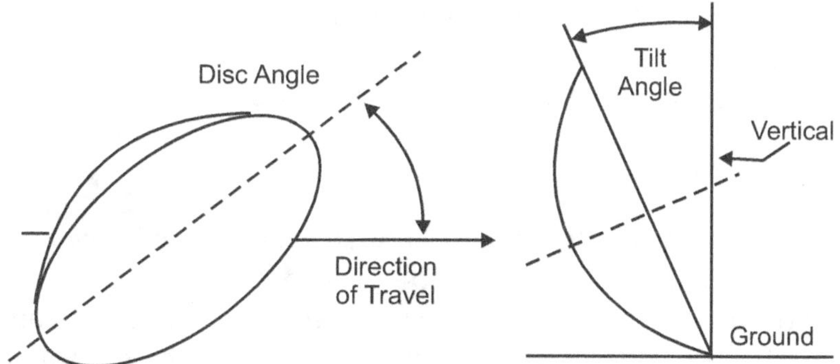


**Fig. 5.24: Scraper**

**Concavity:** It is the depth measured at the centre of the disc by placing its concave side on a flat surface.

## Vertical Disc Plough:

It is the plough which combines the principle of regular disc plough and disc harrow and is used for shallow working in the soil.

### Draft of disc plough:

The disc plough is lighter in draft than the mould board plough, turning same volume of soil in similar conditions. In very hard soil, some extra weight is added to the wheel which increases the draft.

**Rotary Tiller:**

The rotary cultivator is widely considered to be the most important tool as it provides fine degree of pulverization enabling the necessary rapid and intimate mixing of soil besides reduction in traction demanded by the tractor driving wheels due to the ability of the soil working blades to provide some forward thrust to the cultivating outfit. Rotary tiller is directly mounted to the tractor with the help of three point linkage. The power is transmitted from the tractor PTO (Power Take-Off) shaft to a bevel gear box mounted on the top of the unit, through telescopic shaft and universal joint. From the bevel gear box the drive is further transmitted to a power shaft, chain and sprocket transmission system to the rotor. The tynes are fixed to the rotor and the rotor with tynes revolves in the same direction as the tractor wheels. The number of tynes varies from 28 - 54. A levelling board is attached to the rear side of the unit for levelling the tilled soil. A depth control lever with depth wheel provided on either side of the unit ensures proper depth control. The following types of blades are used with the rotor.

(i) **'L' type blade:** Works well in trashy conditions, they are more effective in cutting weeds and they do not pulverize the soil much.

(ii) **Twisted blade:** Suitable for deep tillage in relatively clean ground, but clogging and wrapping of trashes on the tynes and shafts.

(iii) **Straight blade:** Employed on mulchers designed mainly for secondary tillage.

**Chisel Plough:**

Chisel ploughs are used to break through and shatter compacted or otherwise impermeable soil layers. Deep tillage shatters compacted sub soil layers and aids in better infiltration and storage of rainwater in the crop root zone. The improved soil structure also results in better development of root system and the yield of crops and their drought tolerance is also improved. The functional component of the unit include reversible share, tyne (chisel), beam, cross-shaft and top link connection.

**Sub-Soil plough:**

The function of the sub-soiler is to penetrate deeper than the conventional cultivation machinery and break up the layers of the soil, which have become compacted due to the movement of heavy machinery or as a result of continuous ploughing at a constant depth. These compacted areas prevent the natural drainage of the soil and also inhibit the passage of air and nutrients through the soil structure. The sub-soiler consists of heavier tyne than the chisel plough to break through impervious layer shattering the sub-soil to a depth of 45 to 75 cm and requires 60 to 100 hp to operate it. The advantages are same as that of chisel plough.

**Primary and Secondary Tillage:**

Primary tillage is usually conducted after the last harvest, when the soil is wet enough to allow plowing but also allows good traction. Some soil types can be plowed dry. The objective of primary tillage is to attain a reasonable depth of soft soil, incorporate crop residues, kill weeds, and to aerate the soil. Secondary tillage is any subsequent tillage, in order to incorporate fertilizers, reduce the soil to a finer tilth, level the surface, or control weeds.

**Reduced Tillage:**

Reduced tillage or conservation tillage is a practice of minimising soil disturbance and allowing crop residue or stubble to remain on the ground instead of being thrown away or incorporated into the soil. Reduced tillage practices may progress from reducing the number of tillage passes to stopping tillage completely (zero tillage).

Reduced tillage leaves between 15 and 30% crop residue cover on the soil or 500 to 1000 pounds per acre (560 to 1100 kg/ha) of small grain residue during the critical erosion period. This may involve the use of a chisel plow, field cultivators, or other implements. See the general comments below to see how they can affect the amount of residue.

## Intensive Tillage:

Intensive tillage leaves less than 15% crop residue cover or less than 500 pounds per acre (560 kg/ha) of small grain residue. This type of tillage is often referred to as conventional tillage, but as conservational tillage is now more widely used than intensive tillage (in the United States), it is often not appropriate to refer to this type of tillage as conventional. Intensive tillage often involves multiple operations with implements such as a mold board, disk, and/or chisel plow. After this, a finisher with a harrow, rolling basket, and cutter can be used to prepare the seed bed. There are many variations.

## Conservation Tillage:

Conservation tillage leaves at least 30% of crop residue on the soil surface, or at least 1,000 lb/ac (1,100 kg/ha) of small grain residue on the surface during the critical soil erosion period. This slows water movement, which reduces the amount of soil erosion. Additionally, conservation tillage has been found to benefit predatory arthropods that can enhance pest control. Conservation tillage also benefits farmers by reducing fuel consumption and soil compaction. By reducing the number of times the farmer travels over the field, farmers realize significant savings in fuel and labour.

Conservation tillage is used on over 370 million acres, mostly in South America, Oceania and North America. In most years since 1997, conservation tillage was used in US cropland more than intensive or reduced tillage.

However, conservation tillage delays warming of the soil due to the reduction of dark earth exposure to the warmth of the spring sun, thus delaying the planting of the next year's spring crop of corn.

- No-till – Never use a plow, disk, etc. ever again. Aims for 100% ground cover.
- Strip-Till – Narrow strips are tilled where seeds will be planted, leaving the soil in between the rows untilled.
- Mulch-till
- Rotational Tillage – Tilling the soil every two years or less often (every other year, or every third year, etc.).
- Ridge-Till

**Zone tillage:** Zone tillage is a form of modified deep tillage in which only narrow strips are tilled, leaving soil in between the rows untilled. This type of tillage agitates the soil to help reduce soil compaction problems and to improve internal soil drainage. It is designed to only disrupt the soil in a narrow strip directly below the crop row. In comparison to no-till, which relies on the previous year's plant residue to protect the soil and aides in postponement of the warming of the soil and crop growth in Northern climates, zone tillage creates approximately a strip approximately five inches wide that simultaneously breaks up plow pans, assists in warming the soil and helps to prepare a seedbed. When combined with cover crops, zone tillage helps replace lost organic matter, slows the deterioration of the soil, improves soil drainage, increases soil water and nutrient holding capacity, and allows necessary soil organisms to survive. It has been successfully used on farms in the Midwest and West for over 40 years, and is currently used on more than 36% of the U.S. farmland. Some specific states where zone tillage is currently in practice are Pennsylvania, Connecticut, Minnesota, Indiana, Wisconsin, and Illinois. Unfortunately, its use in the Northern Cornbelt states lacks consistent yield results; however, there is still interest in deep tillage within the agriculture industry. In areas that are not well-drained, deep tillage may be used as an alternative to installing more expensive tile drainage.

## Weeding Machines:

The weeds are plants which are considered undesirable in agriculture and gardening. The process of removal of these weeds from crops is called weeding. Weeders are mechanical machines which are used for weed removal.

A weed is essentially any plant which grows where it is unwanted or in the wrong place at the wrong time and doing more harm than good. It is a plant that competes with crops for water, nutrients and light. This can reduce crop production and decrease the value of land, increase cost of cleaning. Weed control is one of the most difficult tasks in agriculture that accounts for a considerable share of the cost involved in agricultural production. Weeding is the removal of unwanted plants in the field crops. Mechanical weed control is very effective as it helps to reduce drudgery involved in manual weeding, it kills the weed and also keeps the soil surface loose ensuring soil aeration and water intake capacity. Farmers generally expressed their concern for effective weed control measures to arrest the growth and propagation of weeds. Chemical method of weed control is more prominent than manual and mechanical methods. However, its adverse effects on the environment are making farmers to consider and accept mechanical methods of weed control. Manual weeding is common in Nigerian agriculture. Today the agricultural sector requires nonchemical weed control that ensures food safety. Consumers demand high quality food products and pay special attention to food safety. These mechanisms contribute significantly to safe food production.

**Types of Weeding Tools:**

The weeding tools and equipment are categorized based on their power source, animal drawn and power or tractor operated.

**Manual weeding tools:**

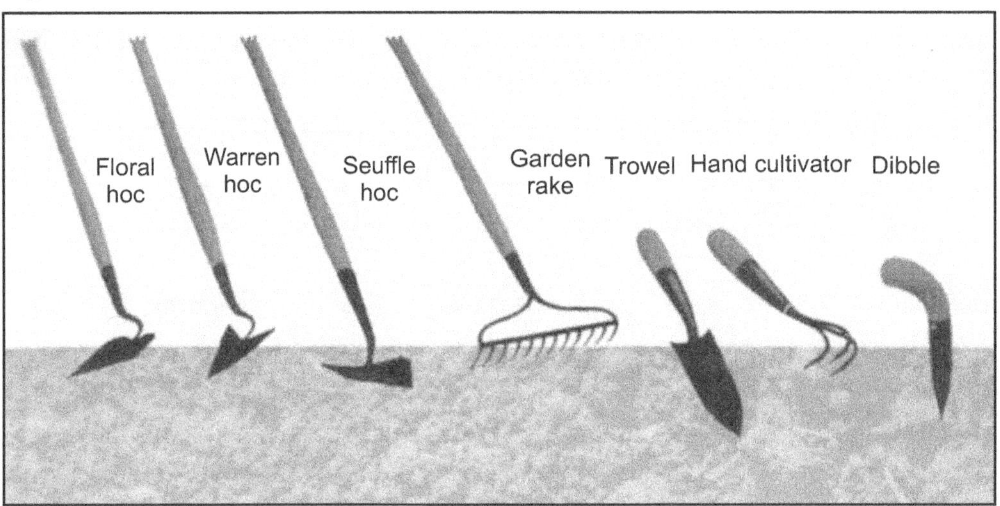


**Fig. 5.25: Manual Weeding Tools**

Manually operated weeders are classified as follows:

(i) **Small tools or aids:** Small weeding tools or aids are traditional hand held type hoes like "Khurpi" used by the farmers. These tools are operated in squatting posture and have very low work output. Different designs of these tools are being used by the farmers of different regions. These tools are suitable for removing the weeds between plants in both row-sown and broadcast fields and are quite efficient.

(ii) **Spades or chopping hoes:** These weeders work on the principle of impact and have straight, curved or pronged blades. Weeds are removed by digging, cutting and uprooting. These are operated in the bending posture. The operation is normally slow and tiring.

(iii) **Long handle tools/weeders:** Long handle tools have a soil working tool mounted at the end of a 1.5 to 2 metre long wooden/bamboo handle. These tools are operated in push or push-pull or pull mode and in standing posture. The soil working tool consists of one or more blades of different shape and size mounted on tines which in turn are fixed on a socket for fitting to the handle. The common shapes of blades on these weeders are straight, convex, V-shape, sweep, serrated etc. These weeders weigh 1.5 to 2.5 "kg. These are designed to work under friable soil moisture conditions and give high work output at the early stages of crop growth when weeds are small.

**Animal drawn weeding tools:**

In oilseed crops, intercultural and weeding operations could be done quickly and efficiently by using improved animal drawn implements. It is essential to provide wider row spacing (above 30 cm), for movement of animals and implement, if animal drawn weeders are to be used. Accurate row spacing and straight rows are a must for successful weeding. Animal drawn tools reduce the cost of operation and time.

**(i)** **Single row hoes:** Animal drawn single row hoes are most widely used by the farmers of different states. Straight or slightly curved blades are commonly used in single row hoes. The size of blade can be changed as per crop row spacing.

**(ii)** **Multi-row implements:** Multi-row units are widely used in Gujarat and other states for wide coverage and timely weeding operation. The three tine cultivator or "Triphali ", "Akola" hoe, "Bardoli" hoe and animal drawn sweeps of different designs are some of the new designs of animal drawn weeders.

**(iii)** **Power operated weeding tools:** Some designs of small engine operated tools have been developed for inter-row cultivation, however, the cost of operation on small farms with a power operated weeder is higher than that with push-pull type weeder. Therefore, their usefulness is limited. Tractor operated implements can be used for intercultivation but these require wider row spacing and leaving of space at the headlands for allowing the tractor to operate and turn before entering into the rows.

**Planter:**

Planter is a machine for placing the seed in continuous flow in furrows at uniform rate and at controlled depth with or without the arrangement of covering them with soil. Seeding methods Common methods used for seeding crop are:

1. Broadcasting
2. Dibbling
3. Drilling
4. Seed dropping behind the plough
5. Transplanting
6. Hill dropping
7. Check row planting

**1. Broadcasting:**

**Fig. 5.26: Broadcasting**

Broadcasting is the process of random scattering of seed on the surface of seed beds. It can done manually or mechanically both. When broadcasting is done manually, uniformity of seed depends upon skill of the man. Soon after broadcasting, the seed are covered by planking or some other device. Usually higher seed rate is obtained in this system. Mechanical broadcasters are used for large scale work. This machine scatters the seed on the surface of the seed bed at controlled rates.

## 2. Dibbling:

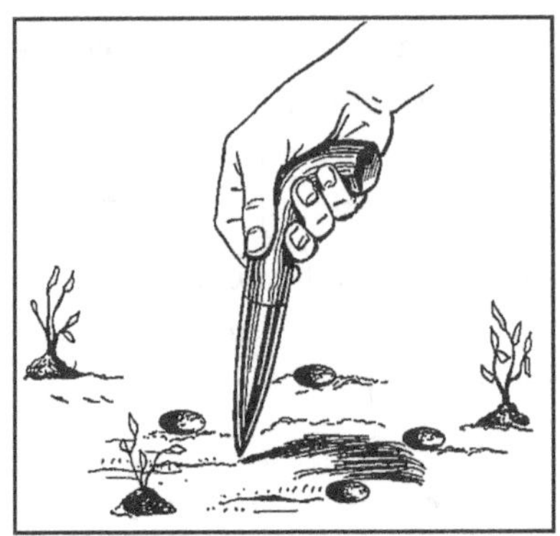

**Fig. 5.27: Dibbling**

Dibbling is the process of placing the seed in hole made in seed bed and covering them.  In this method, seed are placed in holes made at definite depth at fixed spacing. The equipment used for dibbling is called Dibbler. It is a conical instrument used to make proper holes in the field. Small hand dibbler are made with several conical projection made in a frame.

## 3. Drilling:

**Fig. 5.28: Drilling**

Drilling consist of dropping the seed in furrow lines in continuous flow and covering them with soil. This method is very helpful in achieving proper depth, proper spacing and proper amount of seed to be sown in the field.

## 4. Seed dropping:

**Behind the plough:** It is very common method used in villages. It is used for seed like maize, gram, peas, wheat and barley. A man drop seed in furrow behind the plough. Sowing behind the plough can be done by device known as Malobansa. It consist of bamboo tube provide with a funnel shaped mouth.

## 5. Transplanting:

Transplanting consist of preparing seedling in nursery and then planting these seedling in the prepared field. It is commonly done for paddy, vegetables and flowers. It is very time consuming operation.

## 6. Hill dropping:

In this method seed are dropped at fix spacing and not in continuous stream. Thus the spacing between plant to plant in row is constant. In case of drills, the seed are dropped into continuous stream and spacing between plant to plant in a row is not constant.

## 7. Check row planting:

It is method of planting, in which row to row plant to plant distance is uniform. In this method seed are planted precisely along straight parallel furrow. The row is always in two perpendicular directions

The basic objective of sowing operation is to put the seed and fertilizer in rows at desired depth and seed to seed spacing, cover the seeds with soil and provide proper compaction over the seed. The recommended row to row spacing, seed rate, seed to seed spacing and depth of seed placement vary from crop-to-crop and for different agro-climatic conditions to achieve optimum yields.

## Traditional Sowing Methods:

Traditional methods include broadcasting manually, opening furrows by a country plough and dropping seeds by hand, known as 'Kera', and dropping seeds in the furrow through a bamboo/metal funnel attached to a country plough (Pora). For sowing in small areas dibbling i.e., making holes or slits by a stick or tool and dropping seeds by hand, is practised. Multirow traditional seeding devices with manual metering of seeds are quite popular with experienced farmers. Traditional sowing methods have following limitations;

(i)     In manual seeding, it is not possible to achieve uniformity in distribution of seeds. A farmer may sow at desired seed rate but inter-row and intra-row distribution of seeds is likely to be uneven resulting in bunching and gaps in field.

(ii)    Poor control over depth of seed placement.

(iii)   It is necessary to sow at high seed rates and bring the plant population to desired level by thinning.

(iv)    Labour requirement is high because two persons are required for dropping seed and fertilizer.

(v)     The effect of inaccuracies in seed placement on plant stand is greater in case of crops sown under dry farming conditions.

(vi)    During kharif sowing, placement of seeds at uneven depth may result in poor emergence because subsequent rains bring additional soil cover over the seed and affect plant emergence

## Functions of Seed-drills and Planters:

The functions of a well-designed seed drill or planter are as follows:

(i)     Meter seeds of different sizes and shapes;

(ii)    Place the seed in the acceptable pattern of distribution in the field;

(iii)   Place the seed accurately and uniformly at the desired depth in the soil; and

(iv)    Cover the seed and compact the soil around it to enhance germination and emergence.

## Seedbeds for Seeding and Planting:

Depending upon climatic and soil conditions, seeds are sown on well-prepared and levelled fields, on ridges, in furrows or on beds. Flat seeding and planting refer to operation when the field being sown! Planted is levelled and smooth. Seeds and tubers are planted on ridges either to improve soil drainage due to high rainfall or it may be a cropping requirement. Potatoes are usually sown on ridges. Seeding in furrows is done in arid regions to conserve soil moisture and improve plant growth. When two or more rows of seeds are planted in beds and separated by furrows, it is known as bed planting. Bed planting helps in conserving soil moisture, avoids soil compaction and promotes plant growth.

## Subsystems of Sowing and Planting Equipment:

Improved seed-cum-fertilizer drills are provided with seed and fertilizer boxes, metering mechanism, furrow openers, covering devices, frame, ground drive system and controls for variation of seed and fertilizer rates. The major difference in different designs of seed drills/planters is in type of seed and fertilizer metering and furrow openers. Details of these devices are as follows:

## Seed Metering Devices:

Metering mechanism is the heart of sowing machine and its function is to distribute seeds uniformly at the desired application rates. In planters it also controls seed spacing in a row. A seed drill or planter may be required to drop the seeds at rates varying across wide range. Common type of metering devices used on seed drills and planters are:

(i) **Adjustable Orifice with Agitator:** In this type of metering device, seed flow is regulated by changing the size of opening provided at hopper bottom. An agitator fixed above seed opening helps in continuous flow of seeds. This does not give precise control over the seed rate and uniformity of distribution in rows. Many designs of conventional animal and tractor operated machines have adopted this mechanism on account of its simplicity and low cost.

(ii) **Fluted roller (Standard):** Fluted roller metering mechanism is a more positive metering device. Axial or helical flutes are machined or cast on an aluminium, cast iron or plastic roller. Rotation of fluted roller in a housing, filled with seeds, causes the seeds to flow out from roller housing in a continuous stream. Seed rate is controlled by changing exposed length of fluted roller in contact with seeds and fairly accurate seed rate can be achieved for a variety of medium size seeds like wheat, soybean, sunflower and safflower etc. However, metering of small seeds like mustard and sesame at 2 to 5 kg/ha seed rate is not accurate with normal size flutes, designed for seed rates of 20 to 120 kg/ha. Therefore, fluted rollers with smaller size flutes were developed.

(iii) **Fluted roller (Small Flutes):** For sowing of small seeds like rapeseed-mustard and sesame fluted roller with small flutes have been developed at Pantnagar and Ludhiana. These can be fitted by replacing the standard fluted rollers on the seed cum fertilizer drill. The roller is provided with 10 small flutes of 2x2 mm size. Low seed rates of 3 to 5 kg/ha can be achieved with this metering roller with an accuracy of ± 10 per cent.

(iv) **Vertical rotor/Roller with cells:** Vertical rotors with cells are suitable for metering individual or hill of seeds. The rotor with grooves or cells on its periphery is fixed in the hopper. The size and number of cells on the rotor are according to the size of seed and desired seed rate. A cut-off device is provided above the rotor for regulating the flow of seed to cells. In some designs seed rotor is fixed ina secondary hopper and rotor lifts the seeds in cells and drops these into seed funnel. For varying the seed rate and sowing different seeds, separate rotors are required.

(v) **Plate with cells:** Horizontal, inclined or vertical plate with cell type metering mechanism picks and drops individual seed or a hill of seeds depending on design of cell on the plate. Spacing between seeds/hills is controlled by drive ratio ~d number of cells on plate. Separate plates are required for sowing different crops. It is desirable that seeds be graded and have high germination percentage for achieving recommended plant population and uniform seed spacing.

(vi) **Cup feed:** Seed picking cups or spoons are provided on periphery of a vertical plate. When the plate rotates, cups pick seeds from seed hopper and drop them in seed funnel. Size of cups depends on size and number of seeds per hill. This type of metering is used for seeds, which are easily damaged by mechanical devices.

**Furrow Openers:**

The design of furrow openers of seed drills varies to suit the soil conditions of particular region. Most of the seed cum fertilizer drills are provided with pointed tool to form a narrow slit in the soil for seed deposition.

(i) Double end pointed shovel type furrow openers of 100 to 200 room size are used on seed drills in light to medium soils for medium to deep placement of seeds.

(ii) Pointed bar type (diamond shaped) furrow openers are used for forming narrow slit under heavy soils for placement of seeds at medium depths.

(iii) Shoe type openers are used in black soil regions. Seeds are dropped through a tube connected to boot at rear of opener for placement at shallow to medium depths. When used on seed cum fertilizer drills or planters a special narrow boot is designed to place seed and fertilizer in soil at same depth but separated by a small distance.

(iv) Runner or sword type openers are used on planters for shallow sowing. Soil over seed flows back in furrow during operation and seeds are covered with a levelling bar, chain or by operating a wooden plank behind the drill.

## Factors Affecting Seed Germination and Emergence:

Mechanical factors, which affect seed germination and emergence, are:

(i)     Seed damage during metering;

(ii)    Uniformity of depth of placement of seed;

(iii)   Uniformity of distribution of seed along rows;

(iv)    Transverse displacement of seed from the row;

(v)     Prevention of loose soil getting under the seed;

(vi)    Degree of soil compaction above the seed;

(vii)   Uniformity of soil cover over the seed; and

(viii)  Mixing of fertilizer with seed during placement in the furrow.

To achieve the best performance from a seed drill or planter, the above factors are to be optimised by proper design and selection of the components required on the machine to suit the needs of the crops. The seed drill or planter can play an important role in manipulating the physical environment. The metering system selected for the seed should not damage the seed while in operation. The speed of metering device is a very important factor with regards to damage. Seed damage can be avoided by selecting the proper spring loading rate of the cut-off device and knock-out device in case of plate type planters. Seeds should be handled in such a way that physical injury is avoided. Metering devices, seed conveying tubes and their location on the machine affect the seed distribution uniformity in the row. Their selection is affected by the furrow openers and the depth requirements for placement of the seed. The soil physical environment affected by the machine are the availability of soil moisture and mechanical resistance. The planter's performance is improved by manipulating the depth of sowing and thickness of soil cover over the seed as well as pressing the soil cover.

Under arid conditions the top soil becomes very dry, therefore, the seeds are placed 80-100 mm deep in the furrow. This requires the proper furrow opener. Further, the soil cover over the deep-placed seed should be lightly packed to achieve good emergence. In arid regions, dry seeding with deep placement of seed is recommended because the seeds will gem1inateonly when there is sufficient rainfall or adequate moisture at the seed zone.

The recommended fertilizer placement is 50 mm below and 50 mm from the side of the seed. Deep placement of fertilizer is possible if there is no constraint on the power supply to the machine. However, animal-drawn or manually operated planters have this limitation. The compromise is achieved by band placement of fertilizer 40-50 mm from the side of the seed at the same depth.

## Selection of Sowing and Planting Machines:

Different designs of improved seed drills/planters have been developed for sowing of crops. Basic difference in the design of these seed drills is mainly in the type of seed metering mechanism and furrow openers. Therefore, it is essential to select the machine with a metering unit and furrow opener suitable for the crop and soil conditions.

(i)  For small seeds like rapeseed-mustard a seed drill or planter with vertical roller with cells, inclined seed plate with cells or small grooved fluted roller metering system is recommended.

(ii) For medium seeds such as wheat, soybean, safflower and linseed, seed drills with standard fluted rollers are recommended.

(iii) For bold seeds like groundnut and castor planters with inclined cell plate or cup feed type-metering system are recommended.

Furrow openers should be selected according to type of soil and depth of seed placement.

(i) For trashy, stony and light to medium soils, shovel type openers are used. The depth of seed placement from 50 to 100 mm is achieved with these openers.

(ii) Small shoe or shovel type openers are also used for shallow (20 to 50 mm deep) placement of seeds in dry farming areas.

(iii) Shoe type openers with single or twin boots are used for sowing in heavy and medium soils for seed placement at 20 to 70 mm depth.

(iv) Runner type opener is widely used for placement of seeds at shallow depth where soil disturbance required is minimum. Soil cover over seed is also minimal.

Covering chains and wooden planks are widely used to cover and compact the soil over seeds in the furrows and level the fields after sowing operation.

**Harvesting:**

The purpose of harvesting is to recover grains from the field and separate them from the rest of the crop material in a timely manner with minimum grain loss while maintaining highest grain quality. The methods and equipment used for harvesting depend upon the type of grain crop, planting method, and climate. The major grain crops are rice, wheat, corn, soybeans, barley, oats, sorghum, and dry beans (navy) beans, pinto beans, etc.). Many other grain crops, such as oil-seed crops, are harvested using the methods and equipment described. The entire harvesting operation may be divided into cutting, threshing, separation, and cleaning functions. Threshing is breaking grain free from other plant material by applying mechanical force that creates a combination of impact, shear, and/or compression. It is important to avoid damaging grain during threshing–a challenging task under certain crop conditions.

For example, at high moisture content it is harder to break grain away from the crop material but easier to damage grain. The operation of separation refers to separating threshed grains from bulk plant material such as straw. The cleaning operation uses air to separate fine crop material such as chaff from grain.

**Stage of harvest:**

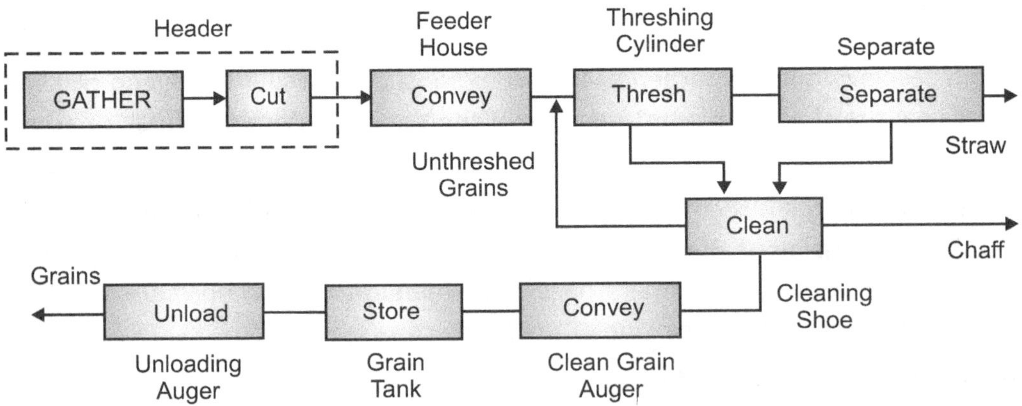


**Fig. 5.29: Stage of Harvest**

Many factors must be considered to obtain optimum rice harvest.

- The grain must be mature, high in quality and have proper moisture content.
- Field should be sufficiently dry to support harvesting and transport.
- Timely harvesting ensures good grain quality, high market value and improved consumer acceptance.

- The right stage for harvesting is when about 80% panicles have 80 % ripened spike lets and their upper portion is straw colored. The grain contains about 20% of moisture.
- Rice should be harvested when the grains on lower part of the panicle are in hard dough stage.
- Maturity may be hastened by 3-4 days by spraying 20 % NaCl a week before harvest to escape monsoon rains

**Harvesting Methods:**

Depending on the size of the operation and the amount of mechanization, rice is either harvested by hand or machine. The different harvesting systems are as follows:

- **Manual harvesting:** Manual harvesting makes use of traditional threshing tools such as threshing racks, simple treadle threshers and animals for trampling or by hand using sharp knives or sickles. Gives 55-60 % grain recovery.
- **Manual harvesting and machine threshing:** Rice is manually threshed, then cleaned with a machine thresher.
- **Machine reaping and machine threshing:** A reaper cuts and lays the crop in a line. Threshing and cleaning can then be performed manually or by machine.
- **Combine harvesting:** The combine harvester combines all operations from paddy harvesting to rice extraction-cutting, handling, threshing and cleaning. Gives 50 % recovery.

**Post Harvesting**

**Threshing:**

- Threshing is the process of beating paddy plants in order to separate the seeds or grains from the straw.
- To maintain the high quality of the harvested grains, it should be threshed immediately after harvesting.
- Avoid field drying and stacking for several days as it affects grain quality due to over drying. Stacked grains of high moisture content results in discoloration or yellowing.
- Threshing can be done manually or mechanically

**Manual threshing:**

- The manual methods of threshing are treading by feet, flail method, and beating stalks against tubs, boards or racks.
- Threshing can be done by trampling using bullocks, rubbing with bare human feet (in hills) or lifting the bundles and striking them on the raised wooden platform.
- Pedal threshers are also used.
- Freshly threshed rice must be dried well in the sun.

**Machine threshing:**

- Mechanical threshing removes rice grains from the rice plant, speeds up threshing (thus reducing losses), and reduces labor requirements.
- Power driven stationary threshers are also used.
- Freshly threshed rice must be dried well in the sun.

**Drying:**

- Drying is the process of removal of excess moisture from the grains.
- Once dried, the rice grain, now called rough rice, is ready for processing.
- Proper drying results in increased storage life of the grains, prevention of deterioration in quality, reduction of biological respiration that leads to quality loss of grains, and optimum milling recovery.

**Methods of drying:**

- Sun drying
- Mechanical drying
- Chemical drying

**Sun Drying:**

- Sun drying is a traditional method of drying the paddy grains. Sun drying is the most economical method of drying grains.
- Grains are spread on drying surfaces such as concrete pavement, mats, and plastic sheets and even on fields to dry naturally.

**Mechanical Drying:**

- Mechanical drying process means drying the grains by ventilating natural or heated air through the grain mass to get it evaporated the moisture from it.
- Mechanical dryers are more reliable since drying could be done anytime of the year.

**Chemical Drying**

- Chemical drying method involves the spraying of common salt solution with specific gravity of 1.1 to 1.2 on the ears of the mature paddy crop.
- This treatment reduces the moisture content from 29% to 14.5% after four days.

**Parboiling**

- Parboiling is a premilling hydrothermal treatment given to unhulled rice (rough rice) to improve its milling quality, nutritive value, cooking quality and storability.
- The process is accomplished in three steps: soaking, steaming and drying.
- Traditionally, parboiling was done either by single or double boiling method.
- Single boiling method involves soaking the unhulled rice in water at room temperature followed by open steaming for 20-30 minutes using iron kettles and then sun drying.
- In double streaming method, the unhusked rice is first steamed to raise its temperature and then soaked in cold water for 24-36 hrs. After soaking, it is steamed second time for 20-30 minutes followed by sun drying.

**Milling**

- Milling is the process wherein the rice grain is transformed into a form suitable for human consumption, therefore, has to be done with utmost care to prevent breakage of the kernel and improve the recovery.
- Brown rice is milled further to create more visually appealing white rice.
- After harvesting and drying, the paddy is subjected to the primary milling operation which includes de-husking as well as the removal of bran layers (polishing) before it is consumed. In this process the rice which is obtained after milling is called raw rice.
- Another process through which rice is obtained after milling is called "Parboiling Rice." Nearly 60% of the total rice produced in India is subjected to parboiling.
- Rice milling losses may be qualitative or quantitative in nature. Quantitative or physical losses are manifested by low milling recovery while low head rice recovery or high percentage of broken kernel reflects the qualitative loss in rice grains.

## Methods of Milling:

**Traditional Method:** Before the advent of mechanical milling, hand-pounding traditional method of rice milling was in practice. In fact, hand-pounding rice has got more nutritive value as compared to machine milling rice. In hand-pounding, a variety of implements is used such as:

- Mortor and Pestle
- Dhenki
- Hand Stone (Chakki)

**Mechanical Method:** With the introduction of mechanized mills, hand-pounding method has steadily decreased because it could not compete with machine mills. The conventional mills in use can be categorized into three main types:

- Huller mills.
- Sheller-Huller mills.
- Sheller-Cone Polisher mills.

### Cleaning and Hulling:

- At the processing plant, the rice is cleaned and hulled.
- At this point, brown rice needs no further processing.
- If white rice is desired, the brown rice is milled to remove the outer bran layers. Hulling is the process to remove the hull from the kernel.

### Manual hulling:

- Hulling can be done by hand by rolling or grinding the rough rice between stones.

### Mechanical hulling:

- The rough rice is first cleaned by passing through a number of sieves that sift out the debris. Blown air removes top matter.
- Once clean, the rice is hulled by a machine that mimics the action of the hand held stones.
- The shelling machine loosens the hulls from the rice. About 80-90% of the kernel hulls are removed during this process.
- From the shelling machine, the grains and hulls are conveyed to a stone reel that aspirates the waste hulls and moves the kernels to a machine that separates the hulled from the unhulled grains.
- By shaking the kernels, the paddy machine forces the heavier unhulled grains to one side of the machine, while the lighter weight rice falls to the other end.
- The unhulled grains are then siphoned to another batch of shelling machines to complete the hulling process.

### Polishing:

- Polishing is the process of removal of bran layer in brown rice.
- After harvesting and drying, the paddy is subjected to the primary milling operation which includes de-husking as well as the removal of bran layers (polishing) before it is consumed.
- The rice obtained after this process is called raw rice.

### Quality and Grading:

- Quality of rice is not always easy to define as it depends on the consumer and the intended end use for the grain.
- Grain quality is not just dependent on the variety of rice, but quality also depends on the crop production environment, harvesting, processing and milling systems.

**Objectives of establishing standards and grades:**

- To ensure only edible rice reaches the consumer.
- To improve post-harvest practices so as to eliminate or reduce waste.
- To improve agronomic practices to increase farm yields.
- To improve processing practices for better milling recoveries and for market expansion.
- To protect consumers from price/quality manipulation.

**Storage:**

Storage is the process of keeping grains, whether in bags or in bulk, in a storage structure designed to protect the stored product from inclement weather and pests for a short or long period of time to await processing or movement to other location.

Grains are stored for either of the following reasons:

- To provide uniform supply of food throughout the year, because grains are produced seasonally while consumption is fairly uniform throughout the year.
- To provide reserve for contingencies such as flood, drought and other calamities.
- To speculate on a good price either in domestic or in the export market.

The grains are stored at three different levels, such as:

- Producer's Level.
- Trader's Level.
- Urban Organizational Storage Level.

The methods followed for storing the grains are:

- Storage in bags.
- Loose storage.
  - Storage in bags is convenient for short term storage, where grain is intended for very early onward movement.
  - For short term storage, no control measures against insects are needed.
  - In loose/bulk storage method, large quantity of grains can be stored in per unit volume of space and the infestation of insects/pests is lower.
  - The basic requirements of a good storage practice are: a healthy, clean, and uniformly dried grain, and a structure that will maintain a suitable environment that will prevent pests.

## 5.4 COLD CHAIN

The term cold chain or cool chain denotes the series of actions and equipment applied to maintain a product within a specified low-temperature range from harvest/production to consumption. A cold chain is a temperature-controlled supply chain. An unbroken cold chain is an uninterrupted series of refrigerated production, storage and distribution activities, along with associated equipment and logistics, which maintain a desired low-temperature range. It is used to preserve and to extend and ensure the shelf life of products, such as fresh agricultural produce, seafood, frozen food, photographic film, chemicals, and pharmaceutical drugs. Such products, during transport and when in transient storage, are sometimes called cool cargo. Unlike other goods or merchandise, cold chain goods are perishable and always en route towards end use or destination, even when held temporarily in cold stores and hence commonly referred to as cargo during its entire logistics cycle.

The cold-chain is a logistical chain of activities involving packaging, storage and distribution of perishable food products (for example, fruits and vegetables, milk, meat and poultry, flowers, and vaccines) from point of production to point of consumption, where the inventory is maintained in predetermined environmental parameters. Typically, a cold-chain is made of 4 links: pack-houses or source point, reefer transport (vehicles or mulita-modal), cold storages, ripening chambers (for some fruits) as illustrated in Fig. 5.30. Refrigeration forms an important and significant part of the food and beverage retail market. It ensures optimal preservation of perishable food. Domestic refrigeration and commercial refrigeration are the important elements of the cold chain.

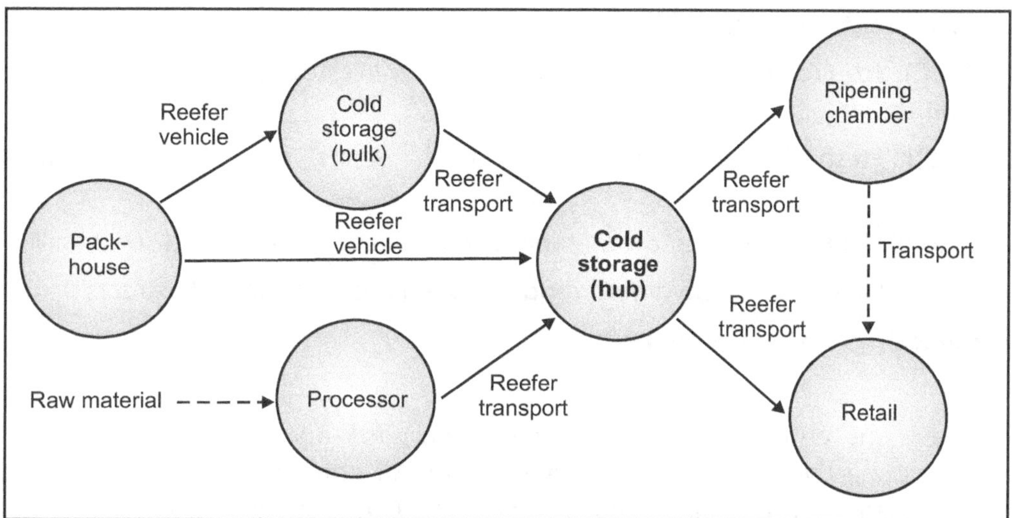


**Fig. 5.30: Schematic Depiction of the Flow of Produce in a Typical Cold-chain**

## 1.   Elements of Cold Chain:

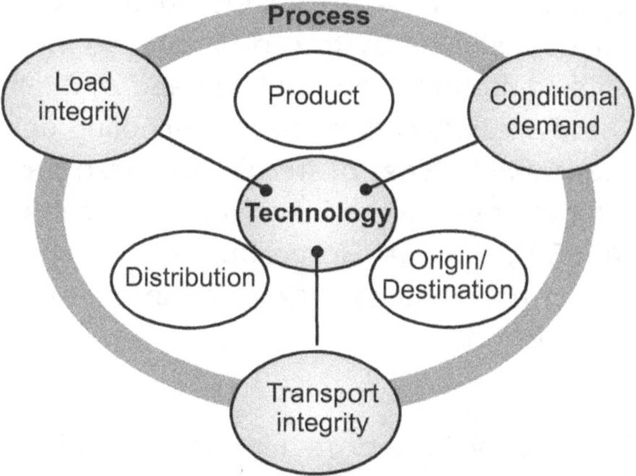


**Fig. 5.31: Elements of Cold Chain**

A **cold chain** is a temperature controlled supply chain. This typically involves keeping items cold from manufacture to point-of-use. This has several elements:

- **Packaging:** Packaging designed to be energy efficient and secure such as insulated shipping containers for fresh sea designed to safely endure short periods of increase temperature as a means of risk reduction.

- **Monitoring:** Monitoring the temperature of items throughout the supply chain with tools such as a temperature data on the cargo, it may also be necessary to monitor other environment parameters such as air quality.

- **Transport:** Cold transport such as refrigerator trucks, refrigerated boxcars, reefer ships and reefer containers.

- **Customs:** A cold chain may pay particular attention to anything that can be done to reduce customs delays. This is process of controlling customs paperwork to ensure it meets all known requirements.

- **Storage:** Cold storage facilities such as temperature controlled warehouses.
- **Quality Assurance:** The process of quality control and managing any quality failures.
- **End-Customer:** Delivery to end-customers and communication of storage requirements. This may involve integration processes for accepting cold deliveries.

## 2. Temperature Requirements of Different Kinds of Produce

### Temperature requirements and Examples:

**Frozen (< −18°C):** Ice cream, frozen meats (fish, poultry, livestock), frozen ingredients, some frozen processed commodities.

**Chilled (0-10°C):** Fresh fruits and vegetables, fresh meats, milk, butter, confectionary, some pharmaceuticals.

**Mild chilled (10-20°C):** Fresh fruits and vegetables, chocolates and seeds and some milk products.

**Normal (> 20°C):** Whole onion, dehydrated foods, pickle, jams and oils and extracts.

## 3. Cold-chain and Refrigeration Components

### Pack-house:

The pack-house is the entry point into the cold-chain. Pack-houses are used to pre-condition the produce for market connectivity and typically serve a range of operations like sorting, grading, washing, drying, weighing, packaging, pre-cooling and staging. The Government has exempted preconditioning services from Goods and Services Tax (GST) to incentivise the development of pack-houses.

The precooling units and staging chambers require energy intensive cooling. Precooling of the produce rapidly for packaging and thereby prepare the loads for subsequent travel in the cold chain. Depending on the produce, the pre-cooling process could be by forced-air cooling, hydrocooling, vacuum cooling, and room cooling or top-icing. A cold staging unit is an insulated and refrigerated chamber which serves as a transient staging space and is a necessary attachment to a pre-cooling unit. A staging room frees the pre-cooler chamber to operate for sequential batches of freshly harvested produce. The country presently has approximately 500 pack-houses (per inputs received from NCCD, Agricultural and Processed Food Products Export Development Authority (APEDA), National Horticulture Board (NHB) and other cold-chain industry experts) but growth in the creation of pack-houses is expected in the next 10-20 years, primarily driven by the Doubling Farmers' Income (DFI) mission.

### Reefer Transport:

A reefer transport unit can be a road vehicle or reefer container, with a fixed insulated body equipped with active refrigeration. There are very small numbers of reefer containers in the country. Over 95% of current existing reefer vehicles are used mainly for carrying frozen products. A typical reefer unit is considered with a holding capacity of 10 MT and is equipped with a refrigeration unit of 3.6-5 kW cooling capacity. However, a variety of sizes are in operation, of smaller and larger capacity. There are presently about 13,000-14,000 reefer vehicles as per the industry information. The number of reefer vehicles is likely to witness a high growth of

20-25% CAGR in the next 10 years while it is expected that the growth in the subsequent decade will be significantly less. Refrigeration systems using CO as a refrigerant and liquid-air based cooling systems are some of the emerging technologies for this sub-sector and are yet to be mainstreamed. Vapour Absorption Refrigeration (VAR) systems are sustainable option wherein waste heat from the prime mover can be utilized.

### Cold Storage:

A typical cold storage facility comprises a highly insulated and refrigerated warehouse designed to store perishable products to essentially maintain the temperature and humidity parameters, which were initiated at the pack-house (for example, vegetables) or during manufacturing (egg, ice-cream).Based on the type of product and holding duration, NCCD has categorized cold storages in two categories, i.e. bulk cold storage and hub cold storage. Bulk cold storage is an environment-controlled warehousing space with multiple chambers intended for bulk storage of perishable produce and is often viewed as an independent refrigerated warehouse, and not part of an integrated cold-chain. The space is designed for long-term storage of a specific produce to build an inventory buffer which will serve to smoothen episodic production by stabilising and sustaining the supply lines. These are normally constructed in areas close to producing areas to facilitate quick access to farmers, for a selective set of crops. Per NCCD, the average holding capacity of a bulk cold storage is about 5000 MT. Majority of these facilities store bulk produce like potatoes and chillies. Hub cold storage (hub), is an environment-controlled warehousing space functioning as a distribution hub located close to consumption centres and form an integral part of the cold-chain. It is designed for short-term handling and cross docking of produce, to serve as a distribution logistics platform at the last mile.  These hubs cold storage provide a platform to manage distribution and delivery activities and are vital to integrated logistics for various products with shorter marketing cycles. The country presently has around 35 million MT of cold storage capacity (as per Directorate of Marketing and Inspection (DMI), Ministry of Food Processing Industries (MoFPI), National Horticulture Mission (NHM), NHB and NCCD), which caters to roughly 85% of the demand.  The existing capacities can also be expected to service larger volume of goods as operations get modernised to optimise on higher rotation of goods from the same space. Cold storage cooling requirements are largely met by vapour compression units using ammonia. The cooling capacity of a cold storage varies between 200-400 kW. A typical cold storage unit runs for 16 to 18 hours per day on grid power. Diesel powered generators are also used across the country and are estimated to meet-up to 10-15% of energy consumption for cold storage operations.

The energy performance of cold storages can be optimised by using improved insulation; variable frequency drives (VFDs), efficient compressors, automation and programmable logic controller (PLC) and retrofitting and retro-commissioning practices.

**Ripening Chamber:**

Ripening chambers are a front-end facility in the cold-chain, used for the controlled and hygienic ripening of certain fresh produce. In India, these are used extensively for ripening bananas and climacteric fruits like mangoes and papayas on commercial scale. The chambers may also be used for avocadoes, tomatoes, pears, and for de-greening purposes with some citrus fruit. A ripening unit usually has 4 compartments, each of around 10 MT capacity. Controlled temperatures of 1520°Cwith elevated humidity levels is typically maintained refrigeration unit. A facility could be in multiples of these units. There are presently about 1000 ripening chambers, which serve around 9% of the current requirement as per assessments carried out by NCCD. Experts predict that there will be a significant growth in ripening chambers (around 9000 units) in the next decade while the subsequent decade would witness a marginal growth. The government has taken special steps to promote modern ripening units so as to eradicate harmful ripening practices, such as use of calcium carbide for ripening.

**Domestic Refrigeration:**

Domestic refrigerators are commonly used in households, commercial setups like retail outlets, offices, hotels, and hospitals for storage of perishable food, medicines, vaccines, etc., These are two types, viz., frost-free (FF) and direct-cool (DC). The production of DC and FF refrigerators has grown at CAGR of 6% and 1% respectively since 2011(BEE) in the country. DC refrigerators dominate the market with around 80% share of the total production from 2011 to 2017 and this trend is likely to continue in the next decade. Based on these historical trends, it is estimated that domestic refrigerator production/sales are expected to at a steady rate of 5-6%. BEE star-rating has been made mandatory for FF refrigerators since 2009 and for DC refrigerators since 2016. The BEE labelling is progressively moving towards higher energy efficiency in this sector.

## Commercial Refrigeration:

Commercial refrigeration equipment covers equipment of different capacities: deep freezers (glass top or hard top) (<1 kW), visi-coolers (<1 kW), remote condensing units (1-20 kW), water coolers (>2 kW), super markets (60-100 kW) and hyper markets (100-200 kW) systems. Remote condenser units could either be display type employed by large retail shops or non-display for storage of additional refrigerated goods. Non-display units have racks of condensing units placed in a small machine room away from the display area. Centralised systems (where compressor racks are installed in a machine room and involve lengthy piping) used in supermarkets and hypermarkets are not very prevalent, but upcoming. The main factors for growth in this commercial refrigeration sector will be commercial space growth, cold chain, GDP growth and technological changes in the future. Per RAMA, the market size of deep freezers, visi-coolers, remote condensing units and waters coolers in 2017-18 is estimated to be around 0.6 million units, 0.3 million units, 0.04 million units and 0.2 million units, respectively.

Described below are two other important extensions of the cold-chain and refrigeration infrastructure in India:

## (a) Cold-chain for vaccine management:

The purpose of the vaccine cold-chain is to maintain product quality from the time of manufacture until the point of administration by ensuring that vaccines are stored and transported within World Health Organisation (WHO)-recommended temperature ranges. The vaccine cold-chain system equipment can be broadly classified under storage and transportation. The storage facilities include walk-in cooler/freezer, deep freezer and ice-lined refrigerator. Solar integrated refrigerant drives have also been a promising vaccine storage facility. Transportation infrastructure consists of refrigerated vaccine van, insulated vaccine van, cold box and vaccine carriers. With the advent of newer technologies in temperature monitoring, cold chain storage and transportation equipment especially using green energy, stock and inventory management, there is a high potential to improve the existing system. India hopes to reach at least 90% full immunization coverage over the next five years, and it is critical to strengthen the vaccine cold-chain in the country. India's Universal Immunization Program is one of the largest in the world and caters to 26 million infants and 30 million pregnant women, saving 2.5 million lives each year. The effectiveness of the program depends largely on a functional end-to-end Immunization Supply Chain system. Immunization supply chain plays a pivotal role in ensuring the uninterrupted availability of quality vaccines from the level of the manufacturer to that of the beneficiary.

## Milk Chillers:

Milk chillers are used in the milk chilling centres and dairy plants. Most of the chilling centres are located in remote villages to collect the milk from various local areas. The transportation of fresh milk from farms to cooling centres and processing units may take sometime. The milk should be chilled within three to four hours of collection otherwise leads to spoilage. Per Balance50Refrigerant Demand in Cold-chain &Refrigeration, the installed capacity of milk chillers was around 0.4 million tonnes of refrigeration (TR) in 2015 and the market is expected to grow at a CAGR of 10%. Ammonia is the widely adopted refrigerant in the dairy sector. Intervention like precooling of warm fresh milk using water will reduce the cooling, refrigerant requirement and energy consumption in the refrigeration system.

## Industrial Refrigeration:

Industrial refrigeration encompasses the cooling systems for production of food, drink, chemicals, pharmaceuticals and other products. This sector also includes systems for controlling air temperature in production factories, computer centres and other process areas. Industrial refrigeration usually uses systems with cooling capacity 10kW to few MW, typically at −50°C to +20°C evaporating temperatures.

## 4. The Future of Cold-chain and Refrigeration:

The waste cold from India's projected LNG imports in 2022 could fuel over half a million liquid air refrigeration units. NCCD is exploring the potential liquid air based cold chains by recovering stranded cold from LNG re-gasification. An analysis - in a report for the NCCD by the energy consultancy E4 tech – shows that a typical LNG terminal re-gasifying 7,100 tons of LNG/day can produce enough liquid nitrogen to provide the cooling for almost 1,100 chilled and frozen refrigerated trucks operating around the clock; and peak time cooling (three hours a day) for 7.5 million cubic meters of chilled and frozen buildings. Dearman, a UK-based technology company, has developed engines that use liquid air/ liquid nitrogen to deliver zero-emission power and cooling. Such engine and cooling systems in reefer transport can be highly efficient with zero polluting emissions. This technology will reduce fuel costs (refrigeration alone consumes as much as 20% of truck's fuel).

Magnetic Refrigeration System (MRS): Magnetic refrigeration is based on the Magneto caloric Effect (MCE). MRS offers the following advantages over compressor-based refrigeration systems:

High Coefficient of Performance (COP) reduces energy consumption by up to 40% No compressor/refrigerant gas used, instead water-based coolant liquid used: eliminates harmful emissions. No gas leakage: reinforced safety and eliminates COOther refrigeration technologies:

- Stirling cycle refrigeration.
- Acoustic refrigeration.
- Electro caloric refrigeration.
- Magnetic refrigeration.
- Optical cooling.
- Thermionic refrigeration.

## 5.5 NATIONAL COOLING ACTION PLAN (NCAP)

Against the backdrop of cooling as a growing developmental need and the international environmental agreements to which India is a signatory, the Ministry of Environment, Forest and Climate Change (MoEF & CC), Government of India,  has taken an initiative to demonstrate a triple-sector  approach to develop and formulate a cooling plan that will resonate with multiple stakeholders in the Government, private sector and non-profit and research organizations in India and help position India as a leader in using sustainable and smart cooling strategies linking Montreal Protocol to both phase-down of HFC's and support Sustainable Development Goals.

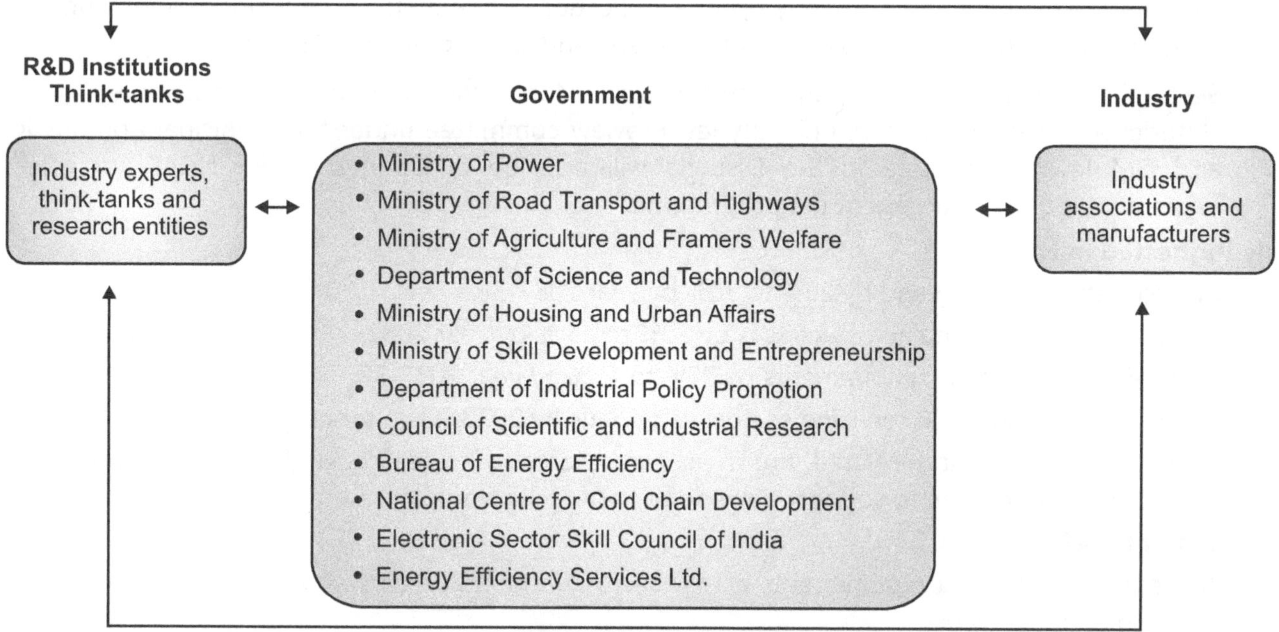


**Fig. 5.32: National Cooling Action Plan Development Framework**

To this end the MoEF and CC decided to develop the India Cooling Action Plan (ICAP) in July 2016 that would provide a 20-year (2017-18 to 2037-38) outlook on how cooling demand in India will evolve and grow. Strategies and actions would be needed in the form of building and habitat design, innovation in the technology (for example, development and launch of air conditioning and refrigerating appliances utilizing low-GWP refrigerants without compromising on energy efficiency, and low energy cooling technologies and constitution of a collaborative research and development (R&D) platform to develop a robust eco-system to promote sustainable and smart cooling strategies while implementing the Kigali Amendment to the Montreal Protocol for Phase-Down of HFC's. The Ministry followed a multi-stakeholder development framework as shown in Fig. 5.32.

## Unique Features of NCAP

1. The overall objective of the programme includes comprehensive mitigation actions for prevention, control and abatement of air pollution besides augmenting the air quality monitoring network across the country and strengthening the awareness and capacity building activities

2. The Environment Ministry has announced a budget of ₹ 300 crore for two years to tackle air pollution across 102 cities, which have been identified by the Central Pollution Control Board (CPCB) for not meeting the pollution standards set by the Ministry of Environment, Forests and Climate Change.

3. Also, city-specific action plans are being formulated for 102 non-attainment cities identified for implementing mitigation actions under NCAP.

4. The Smart Cities programme will be used to launch the NCAP in the 43 smart cities falling in the list of the 102 non-attainment cities.

5. The programme will be institutionalized by respective ministries and will be organized through inter-sectoral groups, which include, Ministry of Road Transport and Highway, Ministry of Petroleum and Natural Gas, Ministry of New and Renewable Energy, Ministry of Heavy Industry, Ministry of Housing and Urban Affairs, Ministry of Agriculture, Ministry of Health, NITI Aayog, CPCB, experts from the industry, academia, and civil society.

6. Other features of NCAP include the increasing number of monitoring stations in the country including rural monitoring stations, technology support, emphasis on awareness and capacity building initiatives, setting up of certification agencies for monitoring equipment, source apportionment studies, emphasis on enforcement, and specific sectoral interventions.

7. Sectoral working groups, national level Project Monitoring Unit, Project Implementation Unit, state-level project monitoring unit, city level review committee under the Municipal Commissioner and DM level Committee in the Districts will also be constituted under NCAP for effective implementation and success of the programme.

## Goals suggested in NCAP:

- Reduce refrigerant demand by 25% to 30% by year 2037-38.
- Reduce cooling demand across sectors by 20% to 25 % by year 2037.
- Reduce cooling energy requirements by 25% to 40% by year 2037-38.
- Train and certify 100,000 servicing sector technicians by 2022-23, in synergy with Skill India Mission.
- Recognize cooling and related areas as thrust area of research under national science and technology programme to support development of technological.

## Objectives of NCAP:

- Assessment of cooling requirements across sectors in next 20 years and the associated refrigerant demand and energy use.
- Map the technologies available to cater the cooling requirement including passive interventions, refrigerant-based technologies, and alternative technologies such as not-in-kind technologies.

- Suggest interventions in each sector to provide for sustainable cooling and thermal comfort for all.
- Focus on skilling of RAC service technicians.
- Develop an R&D innovation ecosystem for indigenous development of alternative technologies.

**Environmental and Socio-economic Benefits:**

- Thermal comfort for all – provision for cooling for Economically Weaker Section (EWS) and Low Income Group (LIG) housing.
- Sustainable cooling – low GHG emissions related to cooling.
- Doubling Farmers Income – better cold chain infrastructure – better value of produce to farmers, less wastage of produce.
- Skilled workforce for better livelihoods and environmental protection.
- Make in India – domestic manufacturing of air-conditioning and related cooling equipment.
- Robust R&D on alternative cooling technologies – to provide a push to innovation in the cooling sector.
- The plan is in sync with India's commitment to the Montreal Protocol, 1987 (reduction of ozone-depleting substances) as well as the Paris Agreement, 2015 to meet the challenges of climate change.

## Multiple Choice Questions

1. The size of a seed drill is expressed by ......
   (a) amount of seed sown per unit time.
   (b) length x width of the machine.
   (c) the number of furrow openers x distance between two furrow openers.
   (d) area covered per unit time.

2. Most suitable furrow opener for seeding In a trashy and hard seedbed is ......
   (a) hoe type          (b) shovel type          (c) disc type          (d) runner type

3. Power operated Paddy transplanter is being manufactured in India by ......
   (a) M/s Escorts Ltd.                    (b) M/s Mitsubishi Ltd.
   (c) M/s Eicher tractors Ltd.            (d) None of the above

4. An indigenous plough is ......
   (a) a multipurpose implement.           (b) a primary tillage implement.
   (c) a secondary tillage implement.      (d) a wetland puddler.

5. The differential lock in a tractor is used to give ......
   (a) high speed to outer wheel on turn.  (b) lower speed to inner wheel on turn.
   (c) different speed to drive.           (d) when two wheels have different traction.

6. Width of cut of disc plough is increased by ......
   (a) Increasing disc angle.              (b) Increasing tilt angle.
   (c) Addition of more weight.            (d) Decreasing disc angle.

7. Soil strength is determined by ......
   (a) Penetrometer.                       (b) Micrometer.
   (c) Hydrometer.                         (d) Dynamometer.

8. The lower value of the ratio of drawbar height to length of tractor is better ......
   (a) variation in weight transfer increased.    (b) variation in weight transfer decreased.
   (c) risk increases while ascending a gradient.  (d) variation in lateral stability is reduced.

9. Soil is derived from Latin word ......
   (a) Solia          (b) Sonum          (c) Solum          (d) Soila

10. Soil erosion is a three phase phenomena ......

    (a) Detachments          (b) Transportation          (c) Deposition          (d) All of these

11. Elements of Cold Chain ......

    (a) Packaging          (b) Monitoring          (c) Transport          (d) All of these

12. Goals suggested in NCAP

    (a) Reduce cooling demand across sectors by 20% to 25 % by year 2037.

    (b) Reduce cooling energy requirements by 25% to 40% by year 2037-38.

    (c) Train and certify 100,000 servicing sector technicians by 2022-23, in synergy with Skill India Mission.

    (d) All of the above.

13. According to India Cooling Action Plan (ICAP), first of its kind, released by Dr. Harsh Vardhan, which of the following year is set as target to reduce cooling demand and refrigerant demand?

    (a) 2021-22          (b) 2027-28          (c) 2037-38          (d) 2022-23

14. Assessment of cooling requirements across sectors in next 20 years and the associated refrigerant demand and energy use.

    (a) True                              (b) False

15. The plan is in sync with India's commitment to the Montreal Protocol is in ......

    (a) 2001          (b) 1992          (c) 1987          (d) 1992

16. Full form of EWS and LIG ......

    (a) Economically Weaker Section and Low Income Group.

    (b) Economically worker Section and Low Income Group.

    (c) Economically Weaker Section and Law Income Group.

    (d) None of these

17. Full form of NCAP is ......

    (a) National cooling action plan.          (b) National calling active plan.

    (c) No cooling action plan.          (d) National cooling action plant.

18. The purpose of harvesting is ......

    (a) recover grains from the field and separate.

    (b) only recover grains from the field and not separated.

    (c) both (a) and (b).

    (d) none of these.

19. The power tillage is most suitable for ......

    (a) Stationary operation.          (b) Deep ploughing.

    (c) Rotary operation.          (d) All are correct.

20. A vertical disc plough is also termed as ......

    (a) Wheat plough.          (b) Harrow plough.          (c) Both (a) and (b).          (d) None of these.

### Answers

| Question No. | 1 | 2 | 3 | 4 | 5 | 6 | 7 | 8 | 9 | 10 |
|---|---|---|---|---|---|---|---|---|---|---|
| Answer | (c) | (c) | (b) | (b) | (a) | (a) | (a) | (b) | (b) | (d) |
| Question No. | 11 | 12 | 13 | 14 | 15 | 16 | 17 | 18 | 19 | 20 |
| Answer | (d) | (d) | (c) | (a) | (c) | (a) | (a) | (a) | (c) | (c) |

✍ ✍ ✍

www.ingramcontent.com/pod-product-compliance
Lightning Source LLC
Chambersburg PA
CBHW081940160726

47999CB00008B/2455